THE MIT PAPERBACK SERIES

A HISTORY OF CIVIL ENGINEERING

A HISTORY OF
CIVIL ENGINEERING

An Outline from Ancient to Modern Times

HANS STRAUB

English translation by
ERWIN ROCKWELL

THE M.I.T. PRESS
Massachusetts Institute of Technology
Cambridge, Massachusetts

Translated from the German
Die Geschichte der Bauingenieurkunst
© Verlag Birkhauser, Basle, 1949

English edition first published 1952
© Leonard Hill Ltd., 1952

First M.I.T. Press Paperback Edition, August, 1964

Copyright © 1964
by
The Massachusetts Institute of Technology
All rights reserved

Library of Congress Catalog Card No. 64-22202
PRINTED IN THE UNITED STATES OF AMERICA

PREFACE TO THE GERMAN-LANGUAGE EDITION

> He who considers things in their first growth
> and origin . . . will obtain the clearest
> view of them. *Aristotle.*[1]

The present book is the outcome of a series of notes which I have compiled, over a number of years, on the historical development of my own profession of Civil Engineering. A long sojourn in Rome, and a keen interest in history at large, filled me with the desire to become familiar with the history of my profession. Social intercourse with members of other professions, particularly architects and art historians, has impressed on me the lack of historical interest on the part of many of my own professional colleagues, and the lack of literature on the subject. He who wants to become acquainted with the origin and development of the art and science of civil engineering, is compelled to resort to a laborious scrutiny of specialist literature on the history of mechanics and statics, and to search old books and scattered contributions to periodicals, difficult to obtain.

My first preoccupation was to form a broad picture of the origin and gradual development of the common conceptions and methods of structural analysis which form the elementary stock-in-trade of the building technician. Similarly, I endeavoured to become familiar with the principal biographical data of those men whose names, though household words to the engineer, have merely come to signify a formula, a relation, or an equation, such as Hooke, Navier, Clapeyron, etc. A series

---

[1] The Politics, Book I, Chapter 2. Benjamin Jowitt's translation; London, 1921.

of contributions to the Swiss journal, *Schweizerische Bauzeitung*, published during the years 1938–1944,[1] are a result of this occasional research. Some of these essays have been embodied in the present book, with little or no change.

My research received an unexpected impetus at the time when, owing to the proximity of the war, engineering work in Rome was practically at a standstill, whilst most of the libraries still remained open. The enforced leisure gave me an opportunity to deepen my studies, and to expand them into a coherent description of the history of civil engineering from the Renaissance to the middle of the nineteenth century. The publishers, Birkhäuser of Basle, to whom the manuscript was submitted, suggested a further revision and expansion so that the book might be embodied in their series of publications, entitled "Science and Culture". In accordance with the tenor of this series, the principal object was to expose the interrelations between the pure science of civil engineering and culture in general, particularly with regard to the different historical styles of art. In other words, a synthesis was to be found between civil engineering proper and culture as expressed by contemporary art. This suggestion was highly welcome as it happened to coincide entirely with my personal intentions and inclinations. It is this further revision which has become the present work.

The book is thus addressed to students and practising engineers as well as to a wider circle of laymen. It is meant to assist the former in viewing their own profession on the broader historical background of science and art, to widen their horizon and to counteract professional narrow-mindedness. Heeding the precept of Aristotle, which has been chosen as a motto for these lines, the academically trained technician should be aware of the origin and development of his profession, of its foundations and roots.

As far as the layman is concerned, the book will, it is hoped, bestow on him an insight into the world of engineering. There is hardly any better way of achieving this object than by an account of the simpler and clearer conditions of the

---

[1] Vol. 112, No. 26; Vol. 116, No. 21; Vol. 118, No. 10; Vol. 119, No. 1; Vol. 120, Nos. 7 and 26; Vol. 123, No. 15.

past when the fundamental principles were less obscured by specialization.

With this dual purpose in mind, the presentation of the subject matter has been kept in a popular vein. The development of the theories of statics and of the strength of materials, as well as their particular application to the requirements of structural engineering is merely sketched in broad outline. Theoretical mechanics are barely touched upon. As far as possible the disquisition of specialized problems has been avoided. Where discussions of a theoretical and mathematical character were unavoidable, e.g., during the historical survey of the origins of structural statics (Sections of Chapter VI; Section 2 of Chapter VIII; and Section I of Chapter IX), these sections may well be omitted by the layman. On the other hand, rather more space has been devoted, even at the risk of repeating well-known facts, to the biographies of those men who have made the greatest contribution to the development of civil engineering. From among the great number of monuments and edifices, only those were selected which can be regarded as typical of the general trend. There are thus a great many important engineering works which have found no mention in this book, and no claim to comprehensiveness is being made.

Mindful of the title of the whole series, I have tried to draw the limits of the subject as wide as possible, and to emphasize the close relationship between civil engineering proper and other spheres of cultural history. In this connection, I have not always tried to suppress my personal opinion. On the other hand, the story has been deliberately confined to *civil* engineering, whilst the development of mechanical engineering and modern industries has only been referred to where this development has a special bearing on problems of civil engineering. It is thus the history of *civil* engineering, and not that of engineering at large, which is being presented.

How modern construction technique and present-day civil engineering have gradually developed from widely divergent roots, namely the science of mechanics on the one hand, and the time-honoured creative craft of building on the other hand — is the picture which the present book is attempting to present.

The two initial chapters which deal with ancient and medieval history are mainly based on information given in various historical works. From the Renaissance onwards, contemporary sources have been used wherever possible, though the use of manuscripts and documentary records had to be renounced. Those sources most frequently resorted to are listed in the bibliography at the end; others are referred to in the footnotes.

HANS STRAUB

PREFACE TO THE ENGLISH-LANGUAGE EDITION

The appearance of an English-language edition of the present work afforded a welcome opportunity of introducing some amendments and corrections, partly based on sources which had not been consulted when the German edition was written. It was found possible to close some gaps, and to follow up a number of useful observations offered by readers.

On the whole, however, the "survey" character of the book has been preserved also in the English edition. Comprehensiveness was not aimed at; it was therefore not feasible to heed all the advice offered by some reviewers. Again, as in the German edition, the main purpose of the book is not the comprehensive presentation of the history of building technique and statical science, but rather the description of the mutual relationship between civil engineering proper and the art of building at large.

I take this opportunity to express my special thanks to the translator, Mr. Erwin Rockwell, for his thorough and conscientious work. So far from being content with tackling, in an exemplary manner, the none-too-easy task of recapturing the spirit of the written word in a different language, he has also, through proposals and advice of his own, helped to round off the book in some places and to eliminate minor errors.

Rome. August, 1951.

HANS STRAUB

CONTENTS

CONTENTS

LIST OF ILLUSTRATIONS

LIST OF ILLUSTRATIONS

LIST OF ILLUSTRATIONS

INTRODUCTION

Since Man has been building, there has been engineering. Ancient engineering works are no less numerous than the works of ancient monumental architecture still in existence. The strength of Roman roads, aqueducts and fortifications still claims our admiration, after 2,000 years. Dimensions and durability of the great vaults of the magnificent halls of the Roman Baths, and of the long flights of stone arches forming the Roman Aqueducts, can well stand comparison with modern engineering works.

There is, however, one fundamental difference between the works of the ancient world and those of modern engineering.

The engineering works of the *ancient* world differ from the contemporary works of monumental architecture merely in their purpose. They are *utilitarian* rather than devotional or monumental in character. There is no clearly defined demarcation between the two; their structure and design is similar in principle; what differs is merely the degree of ornamental embellishment. There are, on the other hand, many masterpieces of monumental architecture (such as the great rotunda of the Pantheon in Rome) which, owing to their gigantic dimensions, may well be regarded as *engineering* structures.

That which distinguishes *modern* engineering works from their contemporary works of monumental architecture is the fact that their dimensions and shapes are determined by theoretical, scientific reflections; that is to say, by abstract devices which are not immediately discernible. The craft of the master builder has been divided into that of the engineer and that of the architect; the one relying primarily on calculation, the other primarily on configuration. There is, of course, no complete divorce; both have much in common. The engineer, and particularly again the modern engineer, who wants to master his craft, must also know how to design; and the architect must be familiar with the methods of structural analysis.

The gradual penetration of the abstract scientific way of thinking into the field of building construction will be the main theme of the second part of this book.

The introduction of the theories of statics and of the strength of materials into building science, which took place during the eighteenth century, may be described, in the words of Albrecht Dürer, as the transition from craft to art in the sphere of engineering construction. Two centuries earlier, this great painter had aspired to a similar aim in the sphere of painting, in an endeavour to raise it from the pursuit of a craft to that of an "art". This, according to his conception, was a complex of precepts, rules and axioms which he evolved, on a *scientific* basis, during his intensive studies of mathematics and of the theory of perspective and proportion. From the dawn of time up to the eighteenth century, the master builder had remained a craftsman who, even in the design of important structures, was mainly guided by intuition. It is well known that the instinctive assessment of statical conditions is largely identical with the artistic appreciation of the structure, so much that we talk of the "instinctive statics of the middle ages". Even at the time of the Baroque, the great masters frequently combined the calling of the formative artist with that of the architect and the engineer. For, these callings differed only in their *tasks;* whereas the mental process of mastering the substance was similar in principle, the difference being merely one of degree. The blend of trained statical intuition with the traditional practical experience of the guilds gave rise to edifices which command our admiration today, even in regard to their statical shape, albeit their conception differs, in principle, from a modern engineering structure.

True, a wealth of mechanical experience had accumulated in the workshops and guilds, in the course of time. In the wake of the Renaissance of the fine arts followed the progress of the technical crafts, mainly in Italy but also in the countries north of the Alps. According to Vasari, the reputation of many contemporary artists was mainly founded on their ability to produce large bronze castings, to bridge large spans or to design ingenious machinery.

However, as Mach[1] remarks, "mechanical experience" must be clearly distinguished from the "science of mechanics" in its modern sense. The primary aim of science is *cognition*. The science of statics, as a branch of theoretical mechanics, is largely a product of modern times, apart from isolated early beginnings (Aristotle, Archimedes, cf. Chapter I, 4; or Jordanus de Nemore in the thirteenth century). The modern science of statics which dates back to, say, the sixteenth century, has developed independent of its practical application and unrelated to the guilds. As we shall see, this development was largely the work of physicists and mathematicians. It is only at a comparatively late stage, during the eighteenth century, that the attempt was made to apply the scientific knowledge gained, and the results of research, to the design of structures and to the solution of practical building problems (cf. Chapters V and VI). The advent of *building statics* as a science, signified the birth of modern structural engineering; it revolutionized the entire art of building; it opened up possibilities previously undreamt of and, in common with the other branches of engineering science, impressed its seal on the nineteenth and twentieth centuries.

If it was stated that development of the science of mechanics was unrelated to the workshops and guilds, this must, of course, not be taken to mean that there existed no contact at all between science and practice. Even Vitruvius demanded that the building artist should be familiar with geometry and arithmetic which, since the dawn of time, were needed for the planning and design of engineering works and buildings, for the regular composition of frontages and layouts, and for the mastery of stone cutting. During the Renaissance, moreover, a knowledge of mathematics and geometry was indispensable for the exact use of perspective, and for the survey and description of ancient ruins. These sciences were therefore eagerly studied by contemporary artists.

The sixteenth century Academy of Arts in Florence, for

---

[1] Mach, p. 1 (see Bibliography at the end).

instance, was a kind of polytechnic college, where the teaching of mathematics was obligatory. Here, mathematics was taught not in its abstract and pure form, but in its purposeful application as the leading science of the art of design ("arti del disegno") which embraced all branches of the technique of arts and engineering.[1] Filled with admiration for the works and teachings of Antiquity, the architects eagerly studied and discussed the scientific and technical writings of Archimedes, Vitruvius and others. In view of the versatility of their interests, it is hardly surprising that men like Leon Battista Alberti and Leonardo da Vinci made valuable contributions to original thought and knowledge, even in the spheres of mathematics and mechanics.

However, with the deepening knowledge of mechanical processes and with the stricter claims on the accuracy of observation and on the irrefutability of proof, scientists tended to specialize, to become *physicists*. True, they were still concerned with time-honoured practical devices such as lever, pulley and tackle, or with ancient problems such as the free fall, the trajectory of missiles, the bending strength of beams and the like. But their primary purpose was to promote research and science, to fathom the laws of nature, rather than to serve the aims of engineering.

[1] cf. Olschki: III, pages 141 and 143.

CHAPTER I

THE ANCIENT WORLD

1. CANALS AND ROADS

Even with the earliest peoples at the dawn of history, the periods of economic and cultural prosperity are closely associated with, and indeed dependent on, the high standard of technical knowledge. The civilizations of the Nile Valley and Mesopotamia which flourished three and two thousand years before our era, could not have arisen without the magnificent canals and irrigation systems which permitted an intensive cultivation of the soil and thus created the living conditions required for a numerous and concentrated population. After the decay of the great irrigation system, Mesopotamia again relapsed into a desolate, non-arable steppe which, even today, is but a thinly-populated living space for roving herdsmen.

Whilst the names and fortunes of famous rulers and warriors have been handed down to posterity, little is known about the builders and organizers of the great hydraulic works and canals of the ancient Egyptians and Babylonians. The small irrigation canals, and the primitive water engines driven by man or ox, may be of casual invention, perfected gradually by generations of peasants. But the major engineering works, such as the ancient canal between the Nile Delta and the Red Sea mentioned by Herodotus[1] and later repeatedly repaired (e.g. under Ptolemy Philadelphus), or the great dams and reservoirs, must have been conceived by single persons or teams of persons of outstanding engineering ability.

In Egypt, the property boundaries had to be fixed afresh

[1] IV/39 and 42.

1

every year, after the great seasonal floods of the Nile. This work called for a certain knowledge of geometry and could only be carried out by trained surveyors. One of the devices used for this purpose was the right-angled triangle with a side ratio of 3 to 4 to 5, thus representing a special case of Pythagoras' theorem. "To what extent the assistance of the surveyor has been sought in the execution of the great engineering works, cannot be proved. . . . It is, however, quite possible that . . . the construction of the great canals was preceded by field surveys."[1]

Although the hydraulic engineers of the stone-less plains of the Nile Delta and Mesopotamia were mainly confined to the use of earth as building material, there are many, and among them important, relics of their dikes and canals in existence, especially in Babylonia.[2]

In Greece and Italy, the dry rocky soil yielded more durable building materials ; but it also confronted the engineers with more varied tasks. There are thus, in Greece, extensive remnants of urban water supply systems, mostly underground ducts, parts of which are still in use today.[3] Some of them were equipped with pressure pipes, e.g. at Pergamon. The Etruscans and Romans were masters of the construction of water supply and drainage works. Witness the ancient Cloaca Maxima in Rome, or the outlets of the Lakes of Albanus and Nemi (built presumably during the fourth century B.C.). The most important work of this kind was the draining of Lake Fucino, under Emperor Claudius, at the time of 40 or 50, A.D. To provide a run-off for the otherwise drain-less lake and its catchment area, a tunnel of $3\frac{1}{2}$ miles length and 28 ft. overall drop was pierced through Monte Salviano to the adjacent Liris Valley. The construction of this tunnel, begun simultaneously from forty shafts, represents an amazing feat of tunnel engineering and survey technique in an age without mechanical aids.[4] The difficulties encountered, mainly due to earth pressure, were

---

[1] Merckel, p. 92 (see Bibliography at the end).
[2] See, e.g., the route map of Mesopotamia shown by Ernst Herzfeld, "Archäologische Reise im Euphrat-und Tigrisgebiet". D. Reimer, Berlin.
[3] cf. E. Curtius, "Die städtischen Wasserbauten der Hellenen". Berlin, 1894.
[4] cf. Cozzo, "Ingegneria Romana". Rome, 1928.

in fact so great that the quality of the work was not uniformly in keeping with its brilliant conception. In spite of the restoration works carried out under Trajan and Hadrian, the tunnel fell into decay during the fifth century. The lake filled up, and it was not before the nineteenth century that it was drained again, namely, under Prince Torlonia, in 1854–1870, when certain sections of the ancient tunnel were used for the new outlet.

That this engineering feat was also the subject of contemporary admiration may be gathered from the fact that it is mentioned by Pliny, Tacitus, Sueton and Dion Cassius.

Whilst this tunnel was built as an aqueduct, there are also early examples of road tunnels, for instance, the so-called "Grotta" of Naples which connects the city proper with the suburb of Bagnoli across the Posillipo Promontory. This tunnel, which is some 2,300 ft. long, dates back to the pre-Christian era. Originally built for pedestrian traffic only, the tunnel has been widened several times in the course of the centuries. Short tunnel sections are also encountered on ancient mountain roads ; for instance, the Petra Pertusa tunnel on the Via Flaminia, constructed under Vespasian.

Another amazing engineering feat of the Romans was the construction of the great aqueducts which, built during the three last centuries B.C. and the first century A.D., provided the capital with an ample supply of fresh water. At the time of Tiberius, nearly 180 million gallons poured daily into Rome. The following table,[1] with particulars of the ten great aqueducts, will serve as a better illustration of the engineering skill of the Romans than any narrative description can do. The names shown in the second column are not those of the actual builder, but those of the official or ruler on whose initiative or under whose auspices the work was carried out.

Apart from hydraulic engineering, the construction of *highways* was among the most important engineering tasks of the ancient peoples. Whilst the highways of empires with centralized government, such as those of the Persians and the Romans,

---

[1] According to Feldhaus ; the same table is quoted by Brunet. For technical details, particularly in regard to the distribution and meter systems, see Kretzschmer, "Rohrberechnung und Strömungsmessung in der altrömischen Wasserversorgung", *Zeitschrift des Vereins Deutscher Ingenieure*, 1934, p. 19.

Year of Construction	Builder	Name	Length	Arcade Length	Cross-sectional Area of Flow
			(in kilometres)		(square metres)
305 B.C.	Appius Claudius	Appia	16.56	0.09	?
263 B.C.	—	Anio vetus	63.60	0.33	0.95
145 B.C.	Marcius	Marcia	91.30	10.25	1.18
127 B.C.	—	Tepula ⎫			0.178 ⎫
35 B.C.	Augustus	Julia ⎭	22.80	9.50	⎬ 0.485
22 B.C.	Agrippa	Virgo	20.85	1.03	1.0
5 B.C.	Augustus	Alsietina	32.80	0.53	variable
—	Augustus	Augusta	1.18	—	?
A.D. 35	Tiberius	Anio novus	86.85	13.0	1.9
A.D. 40–49	Caligula and Claudius	Claudia	68.75	13.0	1.33
			404.69		

were mainly built for strategical and commercial purposes and therefore solely "utilitarian" in character, those of the *Greeks* were venues not only of commerce, but also of devotional processions and pilgrimage, and were built with artistic understanding. In this sphere, too, the Greeks proved to be the muse-inspired people *par excellence.* It was perhaps due to their animistic conception of nature which ascribed a living soul to trees, springs and gorges or regarded them as the abodes of nymphs and demigods, that the Greeks shrank from any violent interference with natural obstacles. The highways were consecrated to the gods. They were decorated with monuments and temples and were, near the towns, lined with tombs. At points of scenic beauty, benches or steps carved into the rock tempted the wayfarer to tarry. While the roadhouses on the Roman and Chinese highways were primarily inns and guest houses, those of the Greeks were dedicated to deities or to the memory of heroes.

On rocky soil, the Greek highways often consisted merely of two wheel ruts, of about 4 ft. 11 in. gauge deliberately carved out or, in some cases, simply worn down in the course of time, thus representing an early form of railway. The artificial ruts normally had a width of about 8 or 9 in., and a depth of 3 to 4 in. which was occasionally increased to as much as

4

12 in. to obtain smooth gradients on irregular, rocky ground.[1] Crossing points for vehicles driven in opposite directions were provided at certain intervals. But far too often, the difficulty of letting each other pass seems to have led to bitter quarrels, such as the one between Oedipus and King Laius which ended in the calamitous patricide that forms the background of Sophocles' tragedy.

The urban streets were mostly paved. Stone-paved market squares are already mentioned by Homer,[2] and excavations have brought to light many remnants of such pavements.

The masters of road engineering, however, were the *Romans*. The barren passes of the Alps, the Pontine Marshes, the steppes and the deserts were no obstacles to their solid highways which enabled their merchants and officials, their dispatch riders and legions to move in safety from one Province to another, at all seasons of the year. Under the Emperors, the Roman Roads extended over some 50,000 miles, and even today, their remnants and traces can be discerned in all parts of the ancient empire, from Spain to Syria, from the Danube to North Africa. Where suitable stone material was available, the old unmetalled commercial routes were superseded by those indestructible highways, paved with heavy polygonal stone flags of amazingly exact pattern, which, even today, bear witness to Roman solidity (Fig. 1).

The construction of the highways in the Provinces was planned, organized and supervised by military "engineers", and carried out by legionaries in times of peace. Forced labour by the rural population or by slaves was, of course, also used for this purpose. The prodigious cost of the great empire highways was defrayed from public funds, or in some cases, from donations by wealthy private individuals. Augustus, for instance, assigned the roads to certain wealthy Senators who were responsible for their maintenance. He himself assumed responsibility for the Via Flaminia.

---

[1] cf. R. J. Forbes, "Notes on the History of Ancient Roads and their Construction". Amsterdam, 1934.

[2] "There, too, is their place of assembly about the fair temple of Poseidon, fitted with huge stones set deep in the earth." (Odyssey VI, V. 266-267, A. T. Murray's translation. London, 1924).

The names of many famous roads, e.g., the Via Appia, are not those of the actual road builder but those of the supervisory official, mostly the Censor, who initiated and supervised the construction. The construction and maintenance of urban streets was generally the duty of the owners of adjacent property.

As regards the technical details of construction, different methods were in use, all according to local conditions. The following method, said to be described by Vitruvius, is frequently mentioned in the literature : On a roadbed of large blocks of stone ("stratumen"), a course of broken stone or debris ("ruderatio") was spread which, in turn, was covered by a layer of sand ("nucleus") and, finally, by the large polygonal basalt blocks ("summum dorsum") with their polished top surfaces which served as the road surface proper. Forbes[1] however, points out that this description (Vitruvius, Book VII, Chapter 1) does not, in fact, apply to the construction of highways but to that of stone or mosaic flooring. In actual practice, this composite cross-section was rarely used for roads. Ashby[2] has found but a single section in Latium which contains all these different courses. Usually, the heavy polygonal basalt blocks were placed in a layer of sand direct on the soil or rock.

In other cases, the top course consisted of tamped broken stone or gravel, resting on a base course of large stone fragments, partly set in lime mortar.[3] In marshy regions, or where stone material was in short supply, the Romans resorted to wooden causeways, resting on pile foundations.

The roadway was mostly lined with kerbstones and, within the villages and towns, flanked by raised footwalks. The engineering works and bridges will be the subject of the next section.

2. BRIDGES AND BUILDINGS

As discussed in the Introduction, it is in the spheres of bridge and building construction that the working method which we have described as typical for engineering methods

---

[1] cf. R. J. Forbes, "Notes on the History of Ancient Roads and their Construction". Amsterdam, 1934.

[2] "The Roman Campagna in Classical Time", 1927, p. 42.

[3] For examples, cf. Merckel, p. 248, and "Handbuch der Architektur," Vol. II.

proper, has been developed first. However, this development is only of fairly recent date ; most of it took place during the eighteenth century. Neither in the Middle Ages nor in the Ancient World do we find signs of, or even hints at, that deliberate, quantitative application of mathematical and physical laws to the determination of dimensions and shape of structures which is the typical feature of civil engineering in its modern sense. True, we can discern a definite trend, from the Egyptians to the time of the Roman Emperors, and once again from the early Romanesque to the late Gothic Period, towards a greater structural "boldness", especially of vaulted structures (using the term "boldness" to express the ratio of the span, and the material employed to bridge it)[1]. The art of vaulting culminated in the halls of the Roman Thermæ as well as in the gothic cathedrals of the twelfth and thirteenth centuries. In both cases, however, the shapes and dimensions were determined merely by what may be called "trained intuition". In spite of the remarkable scientific standard, especially of the Greeks, in the spheres of mechanics and statics (cf. Section 4, page 21) there was hardly any connection between theory and practice, and hardly any attempt to apply the scientific knowledge to practical purposes, in the sense of modern engineering.

We do, however, find indications of deliberate conclusions, drawn from comparatively short experience, in the case of the ancient *Egyptians*, even at a very early stage. The excavations at Sakkara show how, at the time of approximately 3,000 B.C., construction with primitive unburnt bricks was suddenly superseded, partially or wholly, by a new cut stone technique, and how this technique was gradually improved, even within the same monument (Zoser Pyramid), as the treatment of the joints becomes more accurate, and the size of the stone blocks larger.[2]

To the modern engineer, the most conspicuous feature of

[1] In modern engineering, the "boldness" of an arched structure, particularly of a bridge vault, is taken to mean the ratio $1^2/f$ ("Spangenberg's boldness ratio") where 1 is the span and f the rise of the arch.

[2] cf. J.Ph. Lauer, "Fouilles à Saqqarah — La Pyramide à degrés" ; Cairo, 1936. This publication also contains interesting particulars regarding ancient tools found as well as regarding the composition of the mortar used by the ancient builders, consisting of clay, limestone powder and quartz sand.

these oldest buildings still in existence is the almost complete absence of economic principles. The most impressive of these structures, the pyramids (including, first of all the famous pyramids of the Third and Fourth Dynasties, at the beginning of the third millenium B.C.) are nothing but a gigantic stack of stone blocks, shaped to exact geometric patterns and containing, in their interior, a small tomb chamber and one or more narrow access galleries. True, the temples of the later Dynasties (mostly after 1600 B.C.) do contain supporting and space-spanning members. But the tremendous stone columns with the gigantic quasi-architraves, which cover spans of, perhaps, no more than one-and-a-half times the diameter, hardly deserve to be called "structures" from a statical point of view (Fig. 2).

The erection of such structures taxed the knowledge and ingenuity of the Egyptian engineers in much the same way as present-day building tasks call for all the resourcefulness of modern engineers. True, there was then no need for a structural analysis, or for expansion tests and load tests. However, the quarrying of the stone material and its transport from the quarry to the site ; the removal of the enormous blocks (including obelisks of several hundred tons[1]) ; the perfect and indeed, still surviving finish of even the hardest granite ; the organization and catering for the gigantic labour forces[2] required at the edge of the desert, all these problems caused difficulties which can perhaps be properly assessed only by those who are called upon to carry out similar structural tasks in inhospitable regions today, even though they can rely on all the facilities of modern machinery and technique.

---

[1] Also some of the structures erected by the Phœnicians contain gigantic blocks of stone ; the temple terrace at Baalbek, for instance, contains blocks of more than 10,000 cu. ft. (12 x 13 x 64 ft.) ; cf. "Handbuch der Architektur", Vol. II, p. 9.

[2] According to Herodotus (Vol. II, 124), the Pyramid of Cheops, built about 2850 B.C., was built by 100,000 men over a period of 20 years, the teams being relieved every three months. This would correspond to an output of approximately 7½ working days per cubic foot of masonry. This figure may well be exaggerated, though perhaps not as much as might appear at a first glance, having regard to the laborious extraction and treatment of the stone material with the sole aid of primitive tools, and the transport and hoisting of the blocks exclusively by human toil.

Fig. 1. Roman road in the Alban Hills near Rome   *Photo H. Straub*

Fig. 2. Temple of Luxor, Court of Amenophis III

Fig. 3. Ponte Milvio, near Rome

Photo Alinari, Florence

True, the totalitarian rule of the Pharaohs (supported by a loyal caste of functionaries and, at times, priests) permitted the use of unlimited means and slave labour. But even serfs and prisoners must be fed[1] and, in order to carry out the most difficult technical tasks, such as the extraction from bedrock, transport and erection of the large obelisks and of other colossal blocks of stone, an efficient organization of the labour force is indispensable.

The Egyptian structures are the earliest examples of the interrelation between artistic intention on the one hand, and the technical and economic possibilities on the other hand. It is this harmony between technique and style which is typical of the "golden ages" of art, and which we shall have to refer to repeatedly. Lack of statical knowledge calls for simple structures with merely vertical loads. But even with this limitation, the unlimited means available permit the erection of structures of the greatest dimensions and most durable materials. These external conditions react on the formative instinct and determine the artistic taste, creating a type of structure which was still used in Egypt up to the time of the Roman Empire, although by that time, the technical and economic conditions had already undergone considerable changes.

In contrast to Egypt (from the Delta upwards), the plains of Mesopotamia possess very little natural stone material, so that the engineers of Assyria and Babylonia were mainly dependent on *brick* as building material. The Assyrians used to employ unburnt bricks which, in contrast to Egyptian custom, were generally built in whilst they were still moist and were left to dry in the building.[2] The Babylonians of the seventh and sixth centuries B.C., however, normally used burnt bricks, e.g., for the buildings erected by Nebuchadnezzar II. The ruins of the "Tower of Babel," excavated in 1913 by R. Koldewey, show a core of unburnt bricks which is surrounded by a 50 ft. shell of burnt brick. As a binding agent, bitumen was sometimes used instead of mortar.

---

[1] An inscription on a fragment from the time of Rameses III (1200 B.C.) records a strike at the Royal Necropolis as the labourers had been deprived of their rations (Egyptian Museum, Cairo ; Room 29, upper floor).

[2] A. Choisy, "Histoire de l'Architecture". Paris, 1899.

We may now turn to the *Greeks* whose temples, which are their most important buildings, show a definite trend towards a gradually increasing slenderness of the columns and greater intermediate spans, from the Archaic-Doric columns of the temples at Pæstum and in Sicily to the Ionic and Corinthian forms of the fifth and fourth centuries. True, this change of artistic taste is essentially an intellectual process of æsthetic feeling. But the change is unmistakably coinciding with the structural tendency to reach greater heights and to bridge greater spans with the aid of less material. It is therefore not unreasonable to assume that a relationship exists between the gradually emerging preference for more slender and lighter forms, and the growth of a statical feeling which is no longer shrinking back from the construction of higher and more slender supports, and beams of greater spans.

The most important step on the road to a more efficient structural technique, and indeed the prerequisite for the bridging of larger spans, was the invention of *vaulting*. The introduction of the vault into building used to be ascribed to the Etruscans, as the creators of the conservative, sacred and monumental architecture of the Egyptians and Greeks were not familiar with this method of construction. Excavations in Mesopotamia have, however, disclosed much earlier examples (e.g. at Ur, Nuffar and Warka) of underground vaults across canals and tombs, going back to the fourth millenium B.C.[1] The Egyptians, too, used vaulted structures already in the third millenium B.C. though not for important buildings.

The Greeks ascribed the invention of vaulting to *Democritus* (about 470 to 360 B.C.). But, in fact, even the Greeks themselves had already made use of vaults at an earlier time. The technique of vaulting proper was preceded by that of the pseudo-cupolas, such as the one at the Treasury of Atreus at Mycenæ, where a dome above a circular room was constructed by means of successive projecting rings of stone blocks, laid horizontally and becoming progressively smaller towards the top.

The *Romans*, being the classical engineers *par excellence*, brought the art of vaulting proper to an amazing degree of

---

[1] cf. A. Hertwig "Die Geschichte der Gewölbe", *Technikgeschichte*, Vol. 2.

perfection. Even the Etruscans who were, in many respects, the teachers of the Romans, knew how to build vaulted town gateways and bridges, using wedge-shaped blocks without mortar. One surviving example is the 24 ft. span of the bridge at Bieda, north of Rome, built around 500 B.C. A great number of arched bridges, built during the last century B.C. and the first centuries A.D., are still to be seen, either intact or in ruins, throughout the area of the Roman Empire. In Rome alone, there are several of them which are even able to carry the heavy urban traffic of today, though, it is true, after repeated restorations. The most beautiful among them is Hadrian's Pons Aelius, where the three central arches are still intact, and are still showing their original architectural embellishment. Another Roman bridge, Pons Mulvius (Ponte Molle), which carries the heavy traffic of the Via Flaminia across the Tiber about 1½ miles north of the Porta del Popolo, withstood, during World War II, the entire military traffic (including heavy tanks) first of the Italian and German Armies and later of the Allied Forces, the only drawback of the bridge being not lack of strength, but inadequate width (approximately 22 ft. between the parapets). Of the four central arches which are mainly ancient, two are still showing their original structure with archivolts of travertine, whilst the other two have been thoroughly restored during the fifteenth century, the block masonry being partially replaced by brickwork (Fig. 3, Ponte Milvio ; Fig. 4, Ponte Cestio, both in Rome).

One of the most magnificent accomplishments of Roman bridge building was the Nera Viaduct at Narni, built under Augustus and probably destroyed in 1304 by floods of exceptional severity. Of its four arches, with spans varying from approximately 52 to 102 ft., only the one on the extreme left, with a span of 64 ft., has survived.[1] The bridge over the Danube at Turnu Severin, built under Trajan, had a wooden superstructure supported by stone piers.

As is well known, the foundation of river piers confronts the bridge builder with no less a difficulty than the vaulting of the arches. The Roman engineers were familiar with several

[1] cf. G. Albenga, "Il ponte murario romano", in *L'ingegnere*, 1939, p. 869.

methods, including those of pile foundation and caisson foundation, still in use to-day. Vitruvius describes these two methods as follows : "If solid ground cannot be come to, and the ground be loose or marshy, the place must be excavated, cleared, and either alder, olive or oak piles, previously charred, must be driven in with a machine, as close to each other as

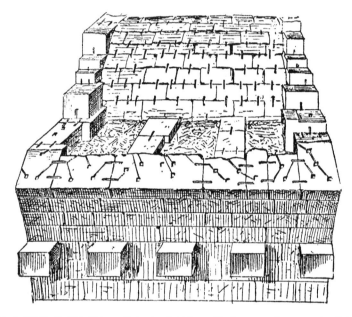

Fig. 4. Ponte Cestio, Rome. Extrados laid bare. To increase the strength of the structure, the blocks are jointed by iron clamps.

From *Annali degli Ingegneri ed Architetti Italiani. Rome*, 1889.

possible, and the intervals between the piles filled with ashes. The heaviest foundation may be laid on such a base."[1] And, as regards the second method : "Then, in the place selected, dams are formed in the water, of oaken piles tied together with chain pieces which are driven firmly into the bottom. Between the ranges of piles, below the level of the water, the bed is dug out and levelled, and the work carried up with stones and

[1] Book III, Chapter 3 (English translation by Joseph Gwilt).

Fig. 5. Aqua Claudia, near Rome

Fig. 6. Four-storey lighthouse and sailing ships. Mosaics in a tomb, Isola Sacra, near Ostia

Photo Dr. H. Jucker

Fig. 7. Pantheon, Rome. Interior, according to Piranesi

mortar, compounded as above directed (viz. hydraulic lime and pozzuolana mortar) till the wall fills the vacant space of the dam."[1]

An important objective of Roman bridge building technique was the erection of canal bridges or aqueducts, needed for the construction of walled-in ducts at a level appropriate for the supply of urban drinking water by gravity. Many miles of arched aqueducts, some of them still standing (Fig. 5), traversed the Roman Campagna. Others, like the Pont du Gard at Nîmes, or the aqueduct of Segovia in Spain, crossed deep valleys in several tiers of arches.

Mention must, finally, be made of the Roman timber bridges, including the legendary Pons Sublicius in Rome, ascribed to the seventh century B.C., and the military bridges, which were erected by engineers of the Roman Army within extremely short periods of time. A prominent example was the Rhine Bridge, constructed by Cæsar's troops in 54 B.C.

It is, however, not only in the sphere of bridge building that the Romans resorted to vaulting ; they also applied it to the construction of *buildings*. The extensive use of the arch is, indeed, a main point of difference between Roman and Greek architecture. Though the Greek columns were retained, they were often degraded to become mere decorative features. For the bearing structure proper, that is to say, for the bridging of spans, the Roman architects preferred the vault. This applies to nearly all monumental building tasks, with the exception of the temples proper. The openings in the frontages of the amphitheatres, basilicas, gateways and triumphal arches, are generally semicircular headed, and the projecting half-columns and architraves have a merely ornamental function. The merely architectural functions of the arch are no concern of civil engineering proper. In contrast, the vaultings of the great interiors of the thermæ, basilicas and other Empire buildings represent supreme achievements of civil engineering : covering, as they do, such spans as the 142 ft. 6 in. span of the Pantheon

---

[1] Book V, Chapter 12. — *Engineering News Record* of 7th December, 1933, p. 675, contains the description of a caisson foundation of limestone for an ancient Egyptian tomb (around 2000 B.C.)

in Rome (Fig. 7) which could only be matched again some fifteen hundred years later, and only exceeded in modern times.

The earlier vaulting technique, using wedge-shaped blocks or concentric rings of radially laid bricks, was replaced, during the first century A.D., by a much more efficient method of vaulting, closely related to modern concrete technique. The vault was erected on a temporary but solid wooden framework with the aid of a kind of casting technique, with alternating courses of mortar and brick or tufa fragments. In the case of major buildings (such as the Pantheon), the casting was preceded by the construction of a system of bearing ribs and relieving arches, in order to reduce the load on the framework.

The soffits of arches, domes and cross-vaults were always semi-circular. Where necessary the thrust of the cross-vaults, concentrated on but a few points, was absorbed by piers and transverse walls which were, however, hardly in evidence from the outside (in contrast to the medieval system of the Gothic Period). In the case of the Basilica of Constantine in Rome (Fig. 12), built in the early fourth century A.D., these transverse walls were included in the building, forming compartments which were covered with semicircular vaults. Above these vaults, the springings of the cross vaults of the main nave, which once covered a span of 83 ft. and have since collapsed, are still visible, together with parts of the supporting piers.

In other cases, e.g. that of the dome which covers the tomb of the Empress Helena, in the vicinity of Rome, an attempt was made to reduce the vault thrust by decreasing the specific gravity of the vaulting material. For this purpose, hollow earthen vessels ("Pignatte") were embedded in the concrete vaulting ; hence the popular name of the monument "Tor Pignattara". This technique was perfected during the fourth and fifth centuries when specially manufactured, wedge-shaped earthen vessels were used, for instance, for several buildings in Ravenna.

The construction of buildings of such dimensions with such enormous vaulted interiors as are evident from numerous relics of the time of the Empire, obviously called not only for advanced technical knowledge but also for a major financial effort. This could only be undertaken by a great power as

14

extensive as the Roman Empire which was, moreover, able to enjoy two centuries of almost uninterrupted peace.[1]

Indeed, there was a profound change very soon after the decay of the Empire. The largest buildings are no longer the vaulted halls of large span ; their place is taken by the Christian basilicas which, though likewise of fairly considerable dimensions, are divided, by rows of columns, into three, or even five naves. The individual spans could be bridged with comparatively light, wooden roof structures which, in contrast to vaulted buildings, do not exert a horizontal thrust. Far thinner walls were therefore sufficient, so that the new construction method permitted a considerable saving in building materials, and thus in cost, even for the same volume of enclosed space. True, the method of vaulting was retained for smaller buildings, baptisteries and the like, and also for certain prominent buildings such as the central domes of S. Vitale, Ravenna, and S. Lorenzo, Milan. Generally, however, the Christian communities of the Occident adopted the system, more appropriate to their financial power, of building timber roof structures, where necessary with intermediate supports to reduce the span.

The history of art records a radical change of the feeling for form and space at the transition from the pagan age of the Romans to the early Christian art.[2] This change of artistic taste is, admittedly, a process pertaining to the spiritual sphere of æsthetics. But once again, we are confronted with a far-reaching coincidence of artistic intention and technical, economic possibilities. There is, however, little doubt that it is the change in external conditions which must be regarded as the primary cause. It is hardly conceivable that this coincidence has come about as a manifestation of providential harmony ; it is much more likely that a virtue has been made of necessity.

3. SHIP BUILDING AND HARBOUR CONSTRUCTION

In ancient times, navigation doubtless had an even greater share in transport at large than is the case in modern times.

---

[1] Concerning the cost of public buildings of the Roman Empire, see Friedländer, "Sittengeschichte Roms", Phaidon Edition, 1934, p. 770.

[2] cf. Samuel Guyer, "Einraum — geteilter Raum", in *Neue Zürcher Zeitung* of 2nd March, 1937, p. 6.

Water-borne transport to-day normally represents but one of several means of goods conveyance, though perhaps the most economic one. In ancient times, however, water-borne traffic often represented the only possibility of conveying bulk goods, and particularly also heavy units, economically, over long distances. Through the invention and development of heat engines during the nineteenth century, overland transport has been revolutionized much more drastically than water-borne transport. The superiority of navigation over other means of transport is therefore less obvious today than it used to be in earlier centuries.

The wooden ships of the Phœnicians, Greeks, Carthaginians and Romans were propelled either by means of sail and wind power, or by means of oars. It is said that the rowers were often arranged in three tiers ("trireme"). It is even alleged that, since the fourth century B.C., vessels with five tiers of rowers have been used.[1] The normal Greek galleys were about 130–165 ft. long and about 16 – 17 ft. wide ; they were manned by approximately 170 rowers. But larger dimensions were not unknown. For instance, the Egyptian grain vessel "Isis" had (according to Merckel) a length of 180 ft., a width of 45 ft. and a height of 43.5 ft. Even larger, it is said, was the famous State barge which was built by Hiero of Syracuse, possibly under the supervision of Archimedes, and presented by him to Ptolemy II of Egypt.

However, the external appearance and the superstructure of the ancient ships cannot be reconstructed with certainty on the basis of traditional descriptions and illustrations (most of which show but *one* tier of oars). We do, however, know some of their structural details from the two State barges of Caligula, raised from the bottom of Lake Nemi, near Rome, some twenty years ago. Unfortunately, these priceless exhibits fell victim to the last war hardly ten years after they had been installed in a museum specially erected for the purpose. Though these ships were flat-bottomed barges rather than seagoing vessels, they were of gigantic dimensions, the larger of them being 233 ft.

---

[1] This is hardly likely to be true, as the top row of oars would have been too long and, thus, too heavy. Even the traditional conception of triremes, with three tiers of rowers at different levels, is open to some doubt.

long and 79 ft. wide. Their destruction must be regarded as an irreplaceable loss, as these vessels were the only surviving specimens of ancient shipbuilding. Apart from their special equipment appropriate to their purpose, the structural features of the Nemi barges may not have differed greatly from those of ancient merchantmen and men-of-war. In fact, even the wooden fishing and sailing boats of today are still being built on much the same principles.

Of particular interest was the equipment of the Nemi barges ; some of it differed amazingly little from modern equipment : two anchors, one of them about 13 ft. long, with a movable anchor stock similar to that of the British Admiralty anchor ; a large bronze cock with extremely accurate fittings ; a primitive piston pump ; a turntable resting on roller supports.

Ancient shipping, relying on comparatively small and weak vessels which were frequently of the open type, called for a great number of not too distant, sheltered harbours, to a greater extent than is the case today. The majority of the Mediterranean harbours which are in existence today were already used in ancient times. This applies particularly to the many natural harbours available along the rugged coastline of Greece. In many cases small islands were connected with the coast by means of a causeway, thus creating two harbours, such as the Portus Aegypticus and Portus Sidonicus at Tyre, or the Portus Major and Portus Minor at Alexandria (Fig. 8).

Just as in the case of road building, there is a distinct difference between the harbour construction works of the Greeks, and those of the more hardened and less sensitive Romans. "Whilst the Greek harbours were always located at places which were predestined for the purpose through the existence of bays or promontories . . . the Romans even selected such sites where everything had to be created by human effort." Wherever possible, the Greeks tried to adapt themselves to the prevailing conditions of soil, current, etc., whereas the Romans did not shrink from radical interference with natural conditions. That is why many of the Roman harbours, though created with enormous effort and with the aid of gigantic engineering works, the relics of which we still admire today, have silted up or have succumbed to the sea : Pozzuoli, Terracina, Anzio,

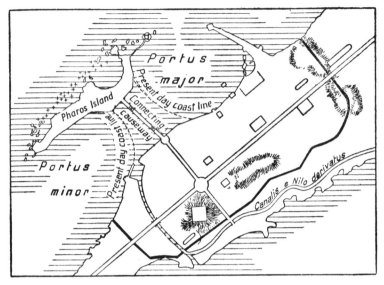

Fig. 8. Port of Alexandria, according to Kiepert-Merckel.

Ostia are examples. Others are still in use today, such as the harbour of Centumcellæ, which was built under Trajan and now forms the inner basin of the modern port of Civitavecchia (Fig. 9). The original harbour, which had remained largely unaltered up to the second half of the nineteenth century, consisted of an inner harbour, now called "Darsena", and a larger basin formed by two tongue-shaped jetties, now called "Molo del Bicchiere" and "Molo Lazzaretto". The entrance to the outer basin was protected by a breakwater, now called "Antemurale Trajano". Unfortunately, this harbour suffered particularly heavy damage during the last war. The two Roman rotundas which flanked the entrance were wrecked, whilst bombs and mines led to the almost complete destruction of the ancient "Darsena". The latter was closely surrounded by fortifications which, though not of Roman origin, dated back to the Renaissance and conveyed a pleasing impression of intimacy of space all the more so as the entrance was no more than about 60 ft. wide and flanked by high buildings. This little harbour basin, teeming with fishing boats, used to

18

present an extremely picturesque aspect, probably not dissimilar to that of an ancient harbour.

At Ostia, the hexagonal harbour basin of 468,000 square yards, built under Trajan, is still in existence today. But, due to the silting up of the coast, caused by the alluvial matter deposited by the Tiber, the ancient harbour now lies two miles inland, forming a picturesque pond surrounded by wall ruins and pines. The scope and magnificent conception of this

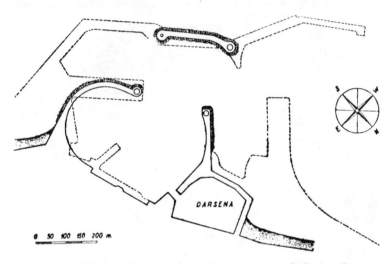

Fig. 9. Harbour of Civitavecchia. Modern parts shown in broken lines.

harbour through which the capital used to receive its supplies may be apparent from the plan (Fig. 10). A reproduction of the four-storied lighthouse is preserved on tomb mosaics on the adjacent Isola Sacra (Fig. 6).

The examination of ancient harbour works is of particular interest, as the construction methods then used have, in many cases, still survived and differ amazingly little from the methods normally used at present.

The substructure of breakwaters and jetties consisted normally of stone ballast. But there are also examples of underwater foundations consisting of rectangular blocks of quarry stone of concrete, which weighed up to 9 tons and were laid out in the shape of regular walls.[1] Pliny describes the construction of the harbour of Centumcellæ : "Huge stones are transported hither in a broad-bottomed vessel, and being sunk one upon the other, are fixed by their own weight, gradually accumulating in the manner, as it were, of a rampart . . ."[2]

Quay walls and smaller jetties were constructed of underwater concrete, in the same way and with the same concrete

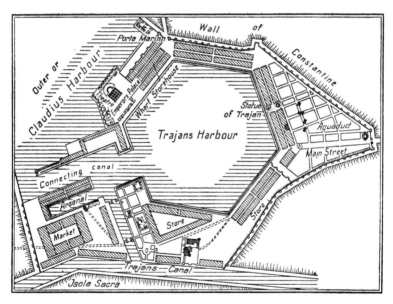

Fig. 10. Trajan's Harbour and Docks near Ostia, according to Lanciani-Merckel.

mix of broken stone, lime and pozzuolana which, in Italy, is still being used for this purpose today. Remnants of hydraulic engineering works, consisting of this extremely tough and durable material, have withstood the physical and chemical aggression of the sea water for nearly two thousand years. They are still to be seen at the coast of Campania and Latium, at Pozzuoli, Formia and Anzio.

[1] cf. *Annali dei Lavori Pubblici*, Rome, 1940, p. 521, and the sources there quoted.
[2] Letter XXXI, William Melmoth's translation. London, 1931.

20

Vitruvius[1] describes a method for the construction of jetties consisting of artificial blocks of pozzuolana concrete. These blocks were cast above water and, after a setting period of two months, lowered into the sea, by causing the sand on which the block was partly resting, to run out as from a sandbox.

4. ANCIENT ENGINEERING SCIENCE — ARCHIMEDES

As already explained (p. 7), there did not exist, during the classical age, an engineering science proper. However, it was the Greeks who, being the creators of scientific thinking in general, laid the foundations of geometry and mechanics, and particularly statics ; that is to say, the very essentials which were later to become the basis of engineering science. In Archimedes of Syracuse arose the world's first great geometrician who combined extraordinary mathematical genius with a high measure of technical skill and insight in problems of mechanics.

The Ionic philosophers of the sixth century B.C., Thales of Miletus and his successors, prepared the ground for the creation of an exact natural science, through their attempt at a monistic and materialistic, rather than supernatural or mystic, explanation of the universe.

Names like Pythagoras, Anaxagoras, Democritus (who proclaimed the atomic composition of matter) and others belong to the history of mathematics and physics, and can here be omitted. But we must occupy ourselves, for a moment, with *Aristotle* of Stagira (384–322 B.C.), whose philosophical and scientific works (including his notes and utterances recorded by the master's disciples and successors, e.g. the famous "Mechanical Problems") had a profound influence on science at large, especially on the mechanics and statics of the Middle Ages and the Renaissance, right up to the sixteenth and seventeenth centuries. It must, however, be admitted that this influence has not been solely to the good. On the contrary, in the field of mechanics, the shadow of the great philosopher constituted, to some extent, a hindrance in the way of progress

---

[1] Book V. Chapter 12.

through many centuries. Only Galilei, replacing traditional belief by his own observation, finally succeeded in doing away with Aristotelian dynamics according to which "heavier bodies were falling more rapidly than lighter bodies" ; "natural motion had to be distinguished from violent motion", and "a thrown body could only continue its flight owing to the simultaneous propulsion of the air", etc.

According to Aristotle, the magnitude of a body-propelling force is expressed by the product of weight (or "mass", which to him is synonymous) and *speed*. Through a reflection which contains the germ of the Principle of Virtual Velocities he arrives at an explanation of the principle of the lever. His "Principle of the Parallelogram of Velocities" is of importance for the subsequent development of structural analysis, as it paved the way for the intuitive recognition of the Principle of the Parallelogram of Forces, once the conception of "force" had been recognized correctly.

Aristotle was the teacher of Alexander the Great, who became the founder of Alexandria, in Egypt. This Greek outpost soon became the intellectual centre of the Greek world, and was provided with research facilities which may well be compared with modern universities and research institutes. Library and "Museion" were soon used by hundreds of scientists and students. Among them was *Euclid*, one of the greatest mathematicians of all times and author of "Elements", that symposium of elementary geometry which, in the logic of its arrangement and in the integrity of its conception, has never yet been excelled. Euclid's methodology which leads with logical force from but a few definitions and axioms to a great range of conclusions, has become the ideal and the unrivalled example of exact science for two thousand years.

Tradition has it that *Archimedes*, too, went to Alexandria, having learned all there was to be learned in his home town, Syracuse, and that he completed his mathematical studies at Alexandria a few years after the death of the author of "Elements".

Archimedes (*circa* 287–212 B.C.) must be regarded as the founder of that physico-mathematical school of thought which in later times (say, since the Renaissance) provoked the amazing development of technical mechanics. He was the most ingenious geometrician of the ancient world, "a genius that can only be compared with Galilei or Newton" (Brunet). His treatise "De Planorum equilibriis" represents the first important work in the special sphere of *statics*, dealing with the theory of the lever, and with the determination of the centres of gravity of such simple figures as the parallelogram, triangle, trapezoid, or the segment of a parabola. According to Brunet,[1] two fundamentally different methods of research must be distinguished in ancient mechanics, the one represented by Aristotle and the other by Archimedes. The former departs from the observation of the individual elementary machines and their working, displaying great skill in formulating the problems, but less felicity in solving them, because his reflections are often perturbed by metaphysical conceptions, such as the assumption that every body is seeking its assigned place.

Archimedes, on the other hand, follows Euclid's method in trying to deduce the principal laws of mechanics, through a logical sequence of thought, from a small number of fundamental principles, open to intuitive conception. Mach describes how the Syracusan proves the general principle of the lever through deducing it from the special case, which he takes for granted, of the symmetric lever (equidistant equal weights are in equilibrium).

The name of Archimedes is best known through his research in the field of hydrostatics. It has been applied to the so-called "Principle of Archimedes" which states that "the upthrust on a body immersed in a fluid equals the weight of the fluid displaced by the body".

Among the achievements of Archimedes in the sphere of pure and applied mathematics may be mentioned his "exhaustion method", the first germ of the Infinitesimal Calculus, which enabled him to calculate the area of circle and ellipse, and the volume of the rotary ellipsoid, paraboloid, etc., as accurately as desired.

But Archimedes is remembered not only as a physicist and mathematician ; he is equally remembered as a technician and

[1] cf. the Bibliography at the end.

engineer of superlative genius. According to Arab sources which cannot be checked, he carried out surveys and built bridges and dams in Egypt. What appears certain is his active interest in applied hydraulics. To him is, for instance, ascribed the invention of "Archimedes' Screw". This device consists of an inclined cylindrical tube containing a spiral which, when immersed in water at its lower end and made to rotate, will raise the water and cause it to flow out at the upper end. It is reported that this device has been used in Egypt for the lifting of irrigation water, in lieu of bucket wheels.

In the preceding section, we have already referred to the construction of the great State ship which is said to have been launched by Archimedes by means of lever and tackle. Whether the inventions of the water-organ and the sundial, the construction of a kind of planetarium, and similar devices are to be ascribed to Archimedes is difficult to decide. But we must regard as fiction the story of the burning mirror which he is alleged to have used to set alight the ships of the Romans who were besieging Syracuse. It is, however, not unlikely that he placed his knowledge of mechanics and his inventive power in the service of the defence of his home town.

There is the well-known legend that Archimedes conceived the idea of his "Principle" while he was taking a bath and then ran home naked, oblivious of his surroundings, and shouting "Eureka". This and other legends bear witness to the deep impression which his intellect and his personality have made on his contemporaries and on later generations. The same can be said of the legends concerning his death : the victorious Roman general wanted to spare the life of the great scientist. But the latter, deeply immersed in studies over his sand table, was slain by a Roman soldier whom he had asked not to disturb his circles.

With Archimedes, the golden age of Greek, and thus of ancient, science had more or less come to an end, though two other great mathematicians were still to follow him, namely Apollonius of Pergamos (between 240 and 170 B.C.), famous for his theory of conic sections, and four centuries later, Diophantus, known as the inventor of algebra.

More closely related to engineering science is the science

of applied mechanics which was mainly at home in Alexandria. True, the water organs, automatic machines and other mechanical and physical toys, constructed by Ctesibios, Philo and Hero, are of greater interest to the mechanical engineer than to the civil engineer. Nevertheless, Vitruvius[1] sees fit to include, among the works which a builder would study with advantage, not only the writings of Archimedes but also those of *Ctesibios*, to whom he ascribes, *inter alia*, the invention of the pressure pump. Of *Philo's* writings, not many have been handed down to us ; apart from treatises of a mechanical character, they also included some on harbour construction, fortifications and catapults, etc.

Most important as a technician and builder was *Hero*, whose exact biographical data are not known. Mach assigns him to the first century B.C., Duhem to the first century A.D. Brunet points out that Hero mentions Euclid, Archimedes and Apollonius, but that Hero himself is only mentioned by Pappus. This would confine the possible life time of Hero to a period from about 150 B.C. to A.D. 250. Of his writings there remains an Arab translation of his "Mechanics", "a complete antique treatise on the theory and application of practical mechanics", and a kind of "engineering manual" which describes, amongst others, the five so-called elementary machines — windlass, lever, pulley (tackle), wedge and worm-gear, as well as several more complicated machines.

Much heeded and often translated during the Renaissance (cf. Chapter III, Section 1) were Hero's writings on war machines and on automatic devices. Best known were his two books on pneumatics which deal with the vacuum ("Hero's Ball"), with the effect of the bent syphon, and the like. A mathematical treatise of Hero's, discovered in Constantinople in 1896 and dealing with the art of surveying, is of special interest in connection with our subject because it conveys, in contrast to the purely scientific character of most of the ancient mathematical disquisitions, an idea of the practical geometry and surveying of the ancient world.

As an example of the ingenious mechanical devices described

---

[1] Book II, Chapter 1.

by Hero in his "Pneumatics" and "Automatons", we may mention his machine (probably only a model) for the automatic opening of the temple gate through the effect of air

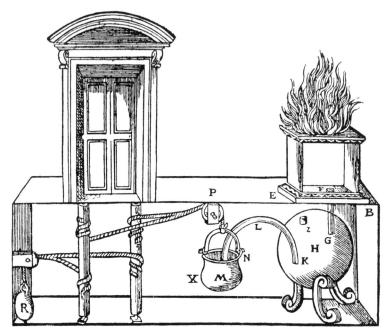

Fig. 11. Hero's machine for the automatic opening of a temple gate. The air, warmed up by the altar fire and expanded, penetrates from the box F into the hollow sphere H which is partly filled with water, and drives part of the water through the pipe L into the cauldron M. The latter, having thus become heavier, is weighted down and, in the process, opens the two sides of the temple gate by means of two ropes which are guided by the pulley P. When the fire is extinguished, the water returns from M into the sphere H, the cauldron becomes lighter, and the counter-weight R becomes strong enough to close the gate again. (According to Brunet and Mieli).

expansion, caused by the heat of the altar fire (cf. Fig. 11 with legend).

The Romans were, as already mentioned, the greatest engineers of the ancient world. As scientists and research workers, however, their achievements do not match those of the Greeks. The extensive writings of *Pliny the Elder* (A.D. 23–79) exercised a profound influence on the natural science of the

26

Fig. 12. Basilica of Constantine, Rome

Fig. 13. Roman hoisting gear : Treadwheel and tackle. Relief in the Lateran Museum, Rome

By courtesy of the Photographical Archives of the Papal Museums

Middle Ages and are still of great value as a means of assessing the knowledge of the ancient world. But Pliny was not much more than an extremely diligent and conscientious compiler. We do, however, possess two very important works dealing with ancient, particularly Roman, engineering, namely Vitruvius' "De architectura libri decem", and Frontinus' "De aquæ ductibus urbis Romæ".

The ten books of *Vitruvius* were written at the time of Emperor Augustus. Little is known about the life and person of the author. The work, which will also occupy us in the following section, represents a complete treatise on the entire art of building, dealing with architecture proper as well as technical problems of construction. The following brief list of the subjects covered by each of the books may convey an idea of the rich contents :

Book 1 : General reflections on the knowledge which an architect and builder requires, and on the subject of architecture in general.

Book 2 : Origin of the art of building ; building materials and their use.

Book 3 : Rules of building design and architecture proper ; construction of temples.

Book 4 : On the design of columns, and further particulars of temple construction.

Book 5 : Other public buildings, especially theatres ; also dealing with harbours and other buildings in water.

Book 6 : Private houses.

Book 7 : On walls, lime, stucco work, painting.

Book 8 : On water supply ; springs, ducts, wells.

Book 9 : On the measurement of time and the different kinds of dials.

Book 10 : On the principles of mechanics (elementary machines such as hoisting gear, scoop wheels and pumps, water organs, catapults and balistæ, etc.).

Frontinus (A.D. 40–103) was a remarkable engineer. As a "curator aquarum", he was charged with the supervision of the aqueducts of Rome, and his work contains an extensive description of this important branch of Roman engineering.

Both Vitruvius and Frontinus were eagerly studied during the Renaissance, and the former exercised a decisive influence on the strictly classicist architectural movement of the mid-sixteenth century, represented by Vignola and Palladio.

5. ENGINEERS, CONTRACTORS AND BUILDING SITES OF THE ANCIENT WORLD[1]

Comparatively little is known about the persons of the ancient builders and engineers. In contrast to the more recent custom of naming exceptional building works after their architects or designers, thus giving them the main credit of the work (we talk, for instance, of Michelangelo's dome of St. Peter's, of Tulla's canalization of the Rhine or of Ammann's Hudson Bridge), the names attached to ancient building works are mostly those of the rulers, war lords or officials, at whose behest the building was erected, or at most, the names of those who placed the order or gave the money for the building. We thus talk of the Appian Way, the bridges and harbours of Trajan, the thermæ of Caracalla and Diocletian. With the Greeks, contempt of the "ignoble" work[2] may have played a part ; with the Romans, the fact that it was the political office only which carried reputation and honour. Even the building supervisors of the Romans were mostly politicians, and not technical experts.

Nevertheless, there have come down to us, from antiquity, a number of names of designers of famous monuments, although the exact share of these men in the execution of the buildings cannot be determined with certainty. It is thus, in many cases, obscure whether they were the designers, or the resident

---

[1] Merckel describes a wealth of details, some of which are used in this section, regarding builders, engineers, building contracts, etc., of the ancient world.

[2] Nevertheless, Herodotus motivates the comparative copiousness of his commentary on the people of Samos by the fact that "they have carried out three works, which are the greatest in all Hellas". Among these are two engineering works proper, namely the great underground aqueduct constructed by Eupalinus, and the great breakwater protecting the harbour. The third of these works is a temple (Herodotus III, 60).

engineers, or the contractors, or simply the supervisory officials in the sense just referred to.[1]

Plutarch, in his "Life of Pericles", names the designers of the great Periclean buildings, including the "Long Walls" built by *Callicrates* which can be regarded as an engineering work proper.

A famous town-planner was *Hippodamus* of Miletus (second half of the fifth century B.C.). The Greeks, who were in the habit of ascribing to a single man developments which, in reality, extended over long periods, ascribed to him the "invention" of regular town plans with rectangular road systems.

At the foundation of the town of Alexandria, Alexander's court-architect, *Dinocrates*, and a mining engineer, *Crates*, are said to have taken part.

Another contemporary of Alexander the Great was *Philo* who is credited with the construction of the arsenal at the Piræus. Around 285 B.C., *Sostratos* built the famous lighthouse on the island of Pharos off Alexandria. This tower,[2] said to have been 443 ft. high, was regarded as one of the seven wonders of the world, and has given its name, in the Latin languages, to lighthouses in general.

Many Greeks used to work in Rome as technicians and architects. Thus, a certain *Hermodorus*, from Salamis, is mentioned as being in charge of the construction of the temples of Jupiter Stator and Mars during the second century B.C. Most famous among the Greek engineers working in Rome is *Apollodorus* of Damascus (early second century A.D.), who worked under Trajan and Hadrian, and is named as the builder of the Forum of Trajan and other public buildings in Rome, as well as of the great bridge over the Danube at Turnu Severin. Under Hadrian, he fell into disgrace and was, it is said, executed.

Many of the architects and engineers working in Rome were, in fact, slaves. The wealthy Crassus, who, together with Cæsar and Pompey, formed the First Triumvirate, is said to

[1] In ancient Egypt, the position of the building supervisor was of high repute, carrying such titles as "The King's Building Supervisor", "Chief of all the King's Building Works", etc. The names of hundreds of such officials have come down to us (Merckel).

[2] Initially, the tower appears to have served merely as a day-time navigation signal, becoming a lighthouse proper only during the first century A.D.

have run a school for the technical training of intelligent young slaves who were then hired out.

Many of the biographical details that have come down to us regarding the activities of ancient engineers, are either scanty or in the nature of legends or anecdotes. A more telling picture can be obtained from a few fragments and inscriptions which we luckily possess, and from several chapters of Vitruvius' books, dealing with technical details. Merckel quotes a report by a public works engineer which is preserved as an inscription, and which conveys a vivid impression of the troubles and surprises in store for the civil engineer, then as to-day. The author of the report, who was apparently the designer and chief inspector, returns to the site of tunnelling work after a long absence : "There I found everybody in a depressed and sulky mood. They had abandoned all hope that the two opposite galleries of the tunnel would meet, as each of them had already been driven more than halfway through the mountain, and no junction had been achieved. As is usual in such a case, the fault was again ascribed to the engineer alone . . . I marked on the mountain ridge the exact position of the tunnel axis. I drew up plans and sections of the entire work . . . I then called the contractor and his men and began the excavation, in their presence, with the help of two shifts of experienced veterans. But during the four years (of my absence) . . . the contractor and his supervisors had committed one error after the other. Each section of tunnel had departed from the straight line, always towards the right-hand side, and if I had come later, Saldæ would have had two tunnels instead of one."

Much in the same way as is done today, the main features and dimensions of ancient building works were determined beforehand by means of more or less accurate project drawings. Fortunately, some of these drawings from ancient Egypt have come down to us ; for instance, that of the rockhewn tomb of Rameses IV, which is preserved on a papyrus kept in the museum of Turin. The accuracy of this plan, which is on a scale of about 1 in 28, has been confirmed in all essentials by recent measurements of the tomb.[1]

[1] cf. "Ägypten und ägyptisches Leben im Altertum" by A. Erman; revised by H.Ranke, Tübingen 1923, page 422. This book contains a reproduction of the plan.

Dörpfeld, basing his assumptions on two inscriptions which have been found, concludes that, in Greece, "in much the same way as today, the technicians had to submit project drawings and descriptions beforehand, and that accurate plans were prepared only after the work was approved. The building programme proper was apparently not accompanied by explanatory drawings. But all the dimensions of the plans were contained in the report".[1]

Vitruvius[2] lists the requirements which a builder and architect should fulfil ; he calls for a good command of the written language, drawing ability, knowledge of geometry, optics and arithmetic. "The architect must have command of the written language so that he can commit to writing his observations and experience in order to assist his memory. Drawing is employed in representing the forms of his design.[3] Geometry affords much aid to the architect : to it he owes the use of the right line and circle, the level and the square, whereby his delineations of buildings on plane surfaces are greatly facilitated. The science of optics enables him to introduce with judgment the requisite quantity of light according to the aspect. Arithmetic estimates the cost and aids in the measurement of the works ; this, assisted by the laws of geometry, determines those abstruse questions wherein the different proportions of some parts to others are involved."

Vitruvius puts forward the rather obvious demand that theoretical knowledge must be matched by practical acquaintance with the craft. "The mere practical architect is not able to assign sufficient reasons for the form he adopts ; and the theoretic architect also fails, grasping the shadow instead of the substance. He who is theoretic as well as practical, is therefore doubly armed, able not only to prove the propriety of his design but equally so to carry it into execution."[4]

Finally, Vitruvius goes so far as to demand, for the architect,

[1] According to Merckel, page 350.
[2] Book I, Chapter 1.
[3] According to Feldhaus (p. 184), the plans were drawn up, by means of a drawing pen, on parchment smoothed by pumice stone.
[4] Quotations from the English translation by Joseph Gwilt.

a knowledge in history, philosophy, music, medicine, law and astronomy. A builder complying with the requirements of the Roman theoretician would thus have to be a master of practically all human knowledge. True, Vitruvius does not fail to give reasons also for these requirements : A knowledge of history is needed to enable the architect to use the proper ornaments. Philosophy bestows on him a certain "nobility of mind" ; moreover, this discipline used to comprise, in the ancient world, the several branches of natural science, particularly mechanics, so that there is perhaps a comparatively valid reason for the inclusion of this subject. A knowledge of music is required to permit an appreciation, on the strength of the sound, of the degree of tension in the ropes of catapults and balistas. Medicine is needed to assess the influence of the climate and of local conditions on the health of the population. Acquaintance with the law is necessary for the construction of eaves, sewers, etc. But this all-too-comprehensive list of subjects which Vitruvius wants his architects to know, tends to arouse a suspicion that this model of an architect is a product of Vitruvius' imagination rather than a reflection of Roman reality.

Generally speaking, the training of the architects and builders of the ancient world was mainly practical.[1] As was the case through the Middle Ages and up to the time of the Baroque, apprentices were taken on by the guilds and on the working sites, and the more capable ones rose to become masters in due course. With few exceptions, their social standing was rather modest, in contrast to that of the Government officials who, as politicians or courtiers, enjoyed a greater prestige, in spite of, or perhaps just because of, their inferior technical knowledge.

Originally, the Government architects and engineers, in Egypt as well as in Rome, apparently belonged to the priesthood, or were at least closely connected with it. This assumption seems to be supported by the use of the Latin word Pontifex, "bridge builder", to denominate a priest. Later on, the con-

---

[1] According to Feldhaus (p. 201), a kind of college for, *inter alia*, mechanics and engineers is said to have been established in Rome under Alexander Severus, about A.D. 228.

struction of temples, aqueducts, roads, etc., was initiated and supervised either, in the case of Greece, by special authorities, or, in the case of Rome, mainly by the Censors who often also gave their names to the buildings. The supervision over public roads was exercised by the ædiles. At the time of the Empire, special "curatores" were placed in charge of the different branches of public works : roads, aqueducts, sewers, water courses.

The work contracts were generally awarded, in the same way as today, to minor or major contractors, according to different methods. The construction of the Long Walls in Athens, for instance, was taken over as a whole by the architect, Callicrates, who assigned the work to sub-contractors in ten allotments. "When the local authorities intend to contract the construction of a temple or the erection of a statue, they interview the artists who apply for the job and submit their estimates and drawings ; whereupon they select the one who, at the lowest price, promises the best and quickest execution."[1] At the Colosseum in Rome, Giuseppe Cozzo[2] noticed slight differences in the construction method of the four sectors and concluded from this fact that the construction had been divided into four allotments which were presumably assigned to different contractors. The building contracts contained, as they do now, detailed specifications regarding the work to be carried out, the materials to be used, the guarantees to be given, and the method of payment.

The building operatives and workers were either free-born or freed, or else slaves. In Rome, the former were organized in brotherhoods or guilds already at an early stage, those of the masons and carpenters being among the oldest. At the time of the first emperors, the slaves, partly recruited from among prisoners of war, formed the great majority of building workers. In fact, the possession of slaves may have represented the bulk of the contractors' stock-in-trade, just as this is the case with machinery and plant to-day.

But even the use of building plant and mechanical devices was not unknown in the ancient world. The Romans, in

[1] Plutarch, as quoted by Friedländer, page 770.
[2] Chapter 4, pages 215 and 224.

particular, sometimes managed to complete major public works within such a short time that the construction would have been impossible without the extensive use of mechanical aids. The Colosseum, for instance, was consecrated but a few years after the construction commenced. Cozzo, however, believes to have found proof for the fact that the supporting structure of travertine, with the staircases and auditorium seats, was completed first whilst the filling walls, consisting of tufa blocks at ground floor level and of brick-faced concrete higher up, were only completed later. Moreover, it would have been impossible to handle, without the aid of hoisting machines, the sometimes enormous components, such as the heavy monolithic columns of marble or granite. One of them, hailing from the Basilica of Constantine and now standing, still intact, in front of the Church of S. Maria Maggiore in Rome, had a length of 47 ft. and a diameter of 5 ft. 9 in. Even greater are the dimensions of the so-called Pompey Column at Alexandria, which consists of Assuan granite ; its shaft has a length of 68 ft. and a diameter of 7 ft. 6 in. at the top, 8 ft. 10 in. at the bottom.

Vitruvius describes, in the Tenth Book of his work, several contemporary building machines, such as pulley, tackle, etc., giving details of their method of operation.

In the Lateran Museum in Rome, there is a relief picture showing an ancient building site, with pulley blocks, windlasses, etc., driven by a large tread-wheel inside which men are walking and climbing. This device has survived, almost unchanged, until modern times, and was finally replaced only by the introduction of the electric motor.[1] The little work of art contrives to present, in a compact and sketchy manner, a vivid picture of an antique building site.

The lively activities on a building site are also, in one instance, described in Roman poetry, namely in the First Book of Virgil's Aeneid where the building of Carthage is thus narrated :[2]

---

[1] The author remembers to have seen, when a boy, such a machine still in operation on a building site in Berne (believed to have been that of the "New Casino").

[2] John Dryden's translation.

34

"The toiling Tyrians on each other call
To ply their labour: some extend the wall,
Some build the citadel; the brawny throng
Or dig, or push unwieldy stones along.
Some for their dwelling choose a spot of ground
Which, first designed, with ditches they surround . . .
Here some design a mole, while others there
Lay deep foundations for a theatre,
From marble quarries mighty columns hew
For ornaments of scenes, and future view."

CHAPTER II

THE MIDDLE AGES

1. THE VAULTING METHODS OF THE ROMANESQUE AND GOTHIC PERIODS.

After the decay of the Roman Empire of the West, the application of vaulting methods in the Occident was, as already mentioned (p. 15), largely confined to small and subordinate buildings.[1] In Byzantium, Asia Minor and Mesopotamia, however, the tradition of dome and vault construction has never lapsed. Witness the Byzantine domes (Hagia Sophia, 532–537, 102 ft. in diameter), the vaulted Basilicas of Asia Minor ; or the Sassanian domes (barrel vault at the palace of Ctesiphon, with 84 ft. span and buttress walls of 23 ft. thickness).

In Europe, it was not before the advent of the Romanesque style in the eleventh century, that the vaulting technique was again applied to major tasks, first in France and somewhat later in Germany and Italy. This development began during the first half of that century, with the construction of barrel vaults over minor single-nave churches in the South-east of France. Later, vaults were also used for the choirs and side aisles, sometimes also for the narrow central nave, of major churches in the Auvergne, in Burgundy and Northern France. There followed the 36 ft. span of the barrel vault, covering the nave of the third church at *Cluny* (begun in 1088, completed in 1131, and demolished in 1811), and the groined intersecting vault of approximately 33 ft. span of the Abbey Church at *Vézelay*, first consecrated in 1104. These churches marked the

---

[1] Among the few exceptions were mainly domes, such as the one above Charlemagne's palace chapel at Aix-la-Chapelle (built in 796-804), or the one above the Baptistery at Florence which has a diameter of 84 ft.

beginning of the grandiose development of the vaulting technique during the next two centuries.

At the same time, or a little later, the first vaulted basilicas were built also in Northern Italy and South-west Germany, among them the cross-vaulted church of S. Ambrogio in *Milan*, with a central nave span of about 40 ft. ; the Abbey of *Laach* in Germany, begun in 1093, which has a central span of about 30 ft., with interesting cross-vaults erected on a rectangular plan ; and particularly the Cathedral of *Spires*. When the construction of this church was commenced in 1030, the building was designed to carry a flat timber roof. It was only under Henry IV, who died in 1106, that the original scheme was modified, and it was decided to use cross-vaults to cover the main nave of 105 ft. height and 43 ft. width. This "mighty prelude to the history of vaulting" represents an achievement which must, however, be ascribed, not so much to novel conceptions of structural analysis, but rather to supreme craftmanship, bold daring and ample funds. As Dehio points out, the decision to cover Spires Cathedral with a vaulted roof was not due to any major progress in engineering science but to the determination of a resolute and powerful king who, it may be relevant to add, also had copious means at his disposal. Although the enormous side walls of over 8 ft. thickness are lightened by means of recesses where the vault thrust is smaller, the quantity of material used was fairly considerable, and economic principles were, as yet, hardly in evidence (Fig. 14).

With the cathedrals built later (Mayence, Worms, and others), little structural progress in vaulting technique was achieved, one of the reasons being that the device of the buttress, long known in France, was disdained in Germany. In the West, however, and in the region of the Isle-de-France, a number of great cathedrals arose in rapid succession during the late twelfth and early thirteenth centuries, which brought the art of vaulting to unprecedented perfection. With the reconstruction, begun in 1137, of the Abbey Church of *St. Denis*, north of Paris, the system now known as "Gothic" was, for the first time, employed consistently. It reached its peak of perfection with the cathedrals of Laon, Chartres, Notre Dame de Paris, Reims, Amiens and others, all commenced between 1155 and 1220 and completed within amazingly short periods (e.g. no more than 25 years in the case of Chartres, and even less in the case of Laon).

As is well known, the structural devices which enabled the Gothic builders to cover a maximum of space with a minimum

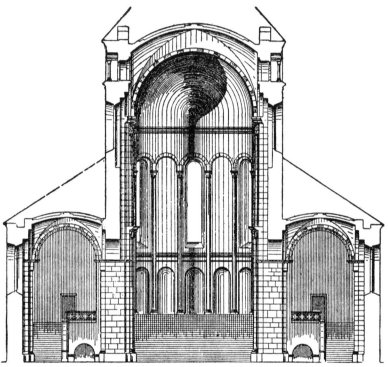

Fig. 14. Spires Cathedral. Cross-section according to Dehio.

quantity of material, and particularly to bestow on the naves of their cathedrals that heaven-aspiring height[1] which so aptly expresses the longing of mystic faith, were these :

1. Distinction between bearing pillars and non-bearing walls, serving merely as enclosure (mostly windows) ;

[1] Amiens Cathedral, for instance, has a vault of 140 ft. height, with a span of 44 ft. The vault of Beauvais Cathedral (which partially collapsed in 1284, shortly after its completion) is even higher, 157½ ft., the width being 51 ft.

2. General application of the easily adaptable pointed arch ;
3. Distinction between vault-supporting ribs and intermediate panels of lighter weight ;
4. Perfection of a system of buttresses and "flying buttresses" to absorb the vault thrust.

We are thus confronted with a kind of framework structure. But, as all the members were merely able to absorb and transfer thrusts, these structures were, of necessity, much more intricate and accomplished than present-day systems which can also make use of tension members and stiff-jointed frames, in addition to pillars and struts.

The consistent application of the principle of concentrating the solid stone material at those points where it had to fulfil a static function, led to considerable saving in material. This will, for instance, be apparent from the right-hand half of Fig. 15, where those parts which would be intersected by a cross-section through the centre line of the windows (i.e. between two successive buttress systems) are shown in black. They are minute in comparison with the buttress system proper (shown in the left-hand half of the drawing) which is, however, of lesser significance as it extends over but a fraction of the length of the building.

The saving in material was, however, partly offset by the greater effort required for the more complicated construction and for the careful, expert treatment of each single stone block. On the other hand, conditions were different from those prevailing today. Whilst wages were low, the quarrying and transport of the stone material was costly owing to the absence of explosives, the poor condition of the roads, and the non-existence of mechanized means of transport.

The pointed arch is statically efficient. It is in greater conformity with the line of thrust than the semi-circular arch which deviates from it considerably at the quarter span. Moreover, as the rise of the pointed arch is greater in relation to the span, the arch thrust is smaller and the thrust line steeper. The break at the point does, however, represent an inconstancy which appears only partly justified, statically, by the weight of the keystone concentrated there.

As regards the vaults, the distinction between bearing ribs

and intermediate panels, sometimes only 6 in. thick, represents a further measure to reduce the weight and, consequently, the thrust, without jeopardizing the safety of the structure.

Fig. 15. Cologne Cathedral. On the left, cross-section through the buttress system; on the right, cross-section through the centre line of the windows. According to Dehio.

The pronounced fashioning of the Gothic vault ribs tends, in spite of the reduced weight, to increase the moment of inertia and thus the buckling strength of these members. The system also tended to facilitate the erection work in as much as a temporary supporting framework was merely required for the ribs, whilst most of the intermediate panels could presumably be laid on the ribs without special temporary supports ("Dovetail vaulting", mainly applied in Germany and England).

But the most convincing manifestation of the professional skill and well developed static feeling of the Gothic architects is the highly accomplished system of buttresses and flying buttresses which they employed to absorb and transmit the arch thrust acting on the springings of the nave vault high above the ground. At these springings, the thrust is divided into two components ; one of them, acting in the vertical direction, is transmitted to the ground through the pier. The other component, pointing outwards in an oblique direction, is transmitted by the "flying buttress" to the next pillar, and possibly to the next but one. In the process, the downwards trend of the resultant oblique pressure is increased through the weight of the pinnacles. There still remains, of course, a horizontal component which is counteracted by the buttress-like enlargement of the outermost pillar (cf. Figs. 15 and 16).

The entire structure forms such a perfectly contrived static system that a modern engineer could hardly improve on it if he were constrained to the use of materials which could only transmit thrust but not tension and bending moments. The degree of perfection is so striking that one art historian who generally prefers to emphasize the æsthetic aspect of architecture, is tempted to suggest that this degree of perfection may, to some extent, already be based on mathematical conceptions.[1]

But the formulas, and the mathematical and geometrical rules of construction in which the professional experience of the guilds was comprehended, and handed down from master to master, were no doubt mostly concerned with questions of

---

[1] "Architecture became a science. To what extent it was, in fact, based on mathematical conceptions, cannot be determined with certainty" (Dehio "Geschichte der Deutschen Kunst", Vol. II, page 29). The opinion that a proper structural analysis has, in fact, been carried out for certain Gothic and, indeed, Byzantine vaults, has also been mooted by Professor A. Hertwig ("Geschichte der Gewölbe", *Technikgeschichte*, Vol. 23, 1934, p. 86). This thesis, however, is not confirmed by factual evidence or by any other sources ; it is merely a subsequent inference, prompted by the purposefulness and perfection of the structures concerned.

form and composition,[1] and had nothing to do with engineering science proper. Another six centuries were to go by before the first attempt was made to design a building element scientifically, on the basis of a structural analysis.[2]

The design of the Gothic vault ribs, buttresses and flying buttresses was thus not based on a proper structural analysis but on experience and intuition. Nevertheless, the general arrangement and design was certainly arrived at on the strength of structural and economic considerations. All the more to be admired is the artistic perfection of all parts. The Gothic architects were never content with the construction as such. Each individual part had a dual function to fulfil, one structural, the other æsthetic, in character. "There is no longer any symbolic use of pseudo-structural forms which were a principal device of Romanesque ornamentality. With strictly realistic truthfulness, each structural member had a definite function to fulfil. Not even the smallest of them could be dispensed with, without bringing down the whole. Structure and decoration have become one."[3]

The French cathedrals of the late twelfth and thirteenth centuries present the most perfect example of that unity of structural technique and artistic appearance, which usually coincides with peak periods of artistic expression, and which has already been referred to in Chapter I. But they also exemplify the fact that such unity is never of long duration. The power of æsthetic expression and the optical attraction of new designs, emanating from structural rather than artistic motives, yet fashioned by real artists, are soon generally perceived and appreciated. As soon as this is the case, the original statical function is often forgotten ; the forms tend to be exaggerated and to be degraded to mere elements of embellishment. This did, in fact, happen soon after the establishment of the system. It found its expression first, in the excessive ornamentality of some cathedrals, and later mainly in the intricacy of the vault ribbing apparent during the late period of English and German Gothic, when the ribs no longer formed the bearing structure of

---

[1] cf. footnote 4 on page 89.
[2] cf. Chapter V, Section 2.
[3] Dehio, "Geschichte der Deutschen Kunst", Vol. I, page 256.

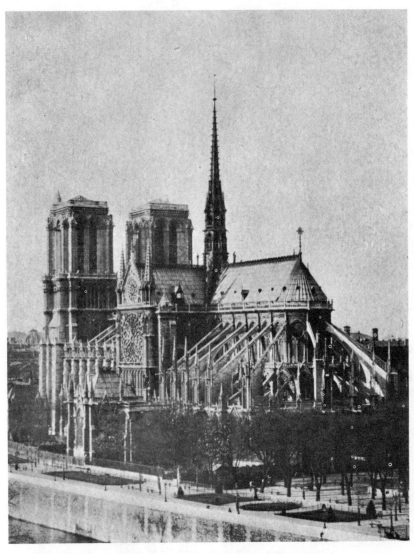

Fig. 16. Gothic Buttress System : Notre Dame, Paris

Fig. 17. Ponte Vecchio, Florence

the vault, but were merely spread over the vault as a decorative feature.[1]

Owing to this romantic blend of artistic and formalistic tendencies and highly developed structural skill, it is hardly appropriate to regard the builders of the great Gothic cathedrals as *engineers* proper. The creators of the French cathedrals, and of the cathedrals of Strasbourg and Cologne, were more than "engineers", and more than "architects" ; they were "*masters*" in the best sense of the word. They were craftsmen, designers and artists all in one who may not have commanded a very extensive knowledge, but who were supreme masters of their profession.

We know little about the personalities of medieval builders, sometimes not even their names. At the time when the first Romanesque churches were built north of the Alps, the monasteries were almost the only abodes of arts and science. We hear, in this connection, sometimes of priests and monks who were also engaged as builders. But they were presumably the exceptions. "For the successful supervision of building work, it is necessary to be steeped in the craft. . . . But the bishops and abbots of those times, being mostly of noble extraction, were fully occupied with their twofold duties of spiritual and temporal regency. Even so, they were still able to exert their influence, to a certain degree, on the design of the building. . . . The dimensions of a church ; the materials to be employed ; the mustering of labour ; the models to be followed, and the innovations to be introduced ; all these problems were the responsibility of the taskmaster. . . . The execution of the work, however, must have been predominantly in the hands of lay-workmen."[2] These may well have been migratory Italian workmen, as the tradition of masonry craftmanship has never died out in the Mediterranean countries since the days of Antiquity. It is, in fact, known that Italian workers were active in Bavaria

[1] In Italy, where the tradition of vaulting has never quite lapsed, large church naves (the width of the central nave of Florence Cathedral, approx. 56 ft., exceeds that of all French cathedrals) had vaults with pointed arches of but little superelevation. As the use of buttresses and flying buttresses was generally disdained, it was often necessary to use ties for the absorption of the vault thrust.

[2] Dehio, "Geschichte der Deutschen Kunst", Vol. I, page 86.

during the eleventh century, and in Saxony during the twelfth century.

Some names have, however, come down to us. We know of one "Magister Odo", cathedral builder of Aix-la-Chapelle, and of the monks, Gauzo and Hezilo, who were in charge of the great reconstruction of Cluny Abbey Church under Abbot Hugo. But most of the clergymen whose names are linked with the building of great churches, such as Otto of Bamberg (Spires Cathedral) or Abbot Suger (Abbey Church of St. Denis), were presumably more concerned with the financial and administrative responsibility than with the technical supervision.

Slightly more is known about building construction during the *Gothic* age. The complicated system of vaults and buttresses called for a high standard of technical knowledge, which could only be attained by trained and experienced specialists. The previously predominant share of the clerical taskmaster in the design and execution of the work waned more and more, as the importance of the guilds waxed. In these, knowledge and experience were accumulated and guarded as professional secrets. Moreover, the work was greatly speeded up through the voluntary services rendered by the local population, particularly in the case of the great French cathedrals : "In that year (1144), for the first time, the faithful of Chartres could be seen hauling the carts which were loaded with stone, timber, corn or anything else needed for the work on the cathedral. As if touched by a magic wand, the turrets rose to the sky. . . . Men and women were seen dragging heavy loads through the marshes and chanting the praise of God's works."[1]

The great masters moved, alone or with a team of skilled workmen, often from one site to another, sometimes over long distances. One of them, Villard de Honnecourt, went from his home town in Northern France all the way to Hungary. His sketch-book has been preserved and is an important source of knowledge concerning the technique of the early thirteenth century. Apart from architectural drawings, it contains structural details of Gothic vaults, a sketch of a kind of truss bridge

---

[1] From the Chronicle of Robert of Mont-Saint-Michel, as quoted by Dehio, "Die Kirchliche Baukunst des Abendlandes", Stuttgart, 1892-1901, Vol. II, page 22.

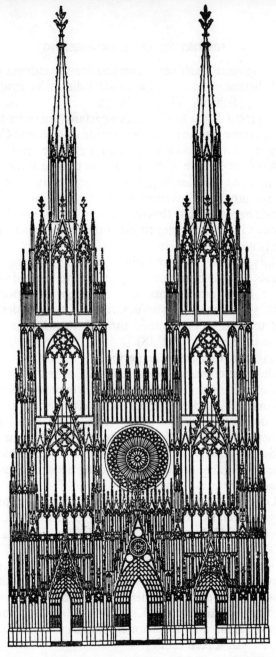

Fig. 18. Gothic design for the main frontage of Strasbourg Cathedral, re-drawn by Dehio.

45

of timber, several machines, a method for the indirect measurement of distances, etc. (cf. Facsimile Edition by Professor H. Hahnloser ; Schroll, Vienna, 1935).

A document dated 1257 names a certain "magister Gerardus Lapicida, rector Fabricæ" as the creator of Cologne Cathedral. Nothing has come down to us about his life. But it can be concluded from certain stylistic features that he must have held a prominent, or even leading, position at the Guild of Amiens.

Several names are linked with the construction of Strasbourg Cathedral, among them that of Master Erwin von Steinbach, familiar to Goethe devotees through the hymn which the young poet dedicated to his memory.

Luckily, some contemporary project drawings of the Cathedrals of Strasbourg and Cologne have come down to us. They are large parchments with frontage elevations which reveal to what extent the structural and formal details of a building used to be determined through drawings prior to the execution of the work (Fig. 18).

During the Romanesque period, "a schematic sketch on which the measurements were entered, as we see it on the drawings for St. Gall, had to suffice to express the wishes of the taskmaster", and all details were left to the artisans. The Gothic drawings, on the other hand, represent complete geometrical project drawings, although these had to undergo repeated changes during the construction period which, as in the case of Strasbourg, extended over several decades and sometimes, centuries.

G. Lenzi[1] gives interesting details of building construction in Italy, especially as regards contract work under King Frederick II of Hohenstauffen, that "first modern man on the throne" (Burckhardt). Even at that time, the lot of the contractor was by no means an easy one. In one case, for instance, an award already made was rescinded when a competitor came forth with a lower offer. Another "prothomagister" whose work did not find favour with the King, was jailed ; his possessions were confiscated, and the former contractor, put in chains, had to perform forced labour.

---

[1] "Il Castello di Melfi e la sua costruzione". Rome, 1935.

46

2. TRANSPORT, ROADS AND BRIDGES OF THE MIDDLE AGES

Compared with Antiquity, the transport conditions of the Middle Ages were decidedly retrograde. Instead of moving on skilfully designed and solidly paved Roman roads, traffic had to negotiate, slowly and laboriously, miserable tracks which were dusty in the summer and mudbound in the winter. In Germany, for instance, road maintenance was confined to the "King's Highways", which were the only routes for through traffic on wheels.

Here and there, the old Roman roads may still have been in use. But, owing to lack of maintenance, their state of repair was often hardly better than that of the unmetalled highways. In fact, the broken remnants of the Roman pavement, torn out and lying about at random, may have been rather a hindrance to wheeled traffic. The numerous surviving Roman *bridges*, on the other hand, continued to fulfil an important function in bridging the rivers. For, the bridge building achievements of the early Middle Ages, up to the end of the eleventh century, were mainly confined to the erection of a few wooden trestle bridges, ever exposed to destruction by fire or floods. Minor watercourses were mostly forded. There is, for instance, a stipulation in the Carolingian "Capitulare de villis vel curtis Imperii", prescribing the use of waterproof leather covers for wagons "so that no freight may suffer damage from rain or from the fording of rivers".

A number of stipulations concerning highways and highway traffic are contained in the "Sachsenspiegel", the oldest German code of law which was compiled between 1198 and 1235 : ". . . The highways must be so wide that wagons can pass each other. The pedestrian must give way to the horseman ; the horseman to the wagon, the empty wagon to the loaded wagon."[1]

Later during the Middle Ages, some progress in the field of transport is beginning to become apparent. A prominent case in point is the opening-up, in the early thirteenth century, of the St. Gotthard Pass, the shortest and most direct north-

---

[1] Quoted by Feldhaus, page 283.

south route across the Alps. Up to that time, the forbidding defile of the Schöllenen Gorge had prevented any direct exchange of goods between the Urseren Valley and the lower valley of the Reuss. The opening-up of the defile by means of a properly maintained pack-road and several bridges, some of them suspended from the rock by chains[1] (e.g. the well-known "Stiebende Brücke"), represented a feat of the first order, even though it can hardly be regarded as an engineering work proper. The creation of the new north to south connection, which also comprehended the waterways of the Lake of Lucerne and the Lago Maggiore, signified a great step forward in the transport conditions of Central Europe.

In the lowlands, the improvements in transport facilities attained during the twelfth and thirteenth centuries consisted mainly in the construction of bridges, many of them solid stone bridges which took the place of earlier wooden bridges. Among the bridges built during the twelfth century were those over the Danube at Ratisbon (1135–1146), over the Elbe at Dresden, over the Main at Würzburg, over the Moldau in Prague, as well as the oldest stone bridges over the Thames in London (London Bridge, demolished in 1831) and over the Arno at Florence (at the site of the present Ponte Vecchio). More in line with Roman tradition (having, for instance, flood outlets above the piers) is the famous Rhone bridge of Avignon, built 1177–1185, four arches of which are still standing. The construction of this bridge is ascribed to St. Bénézet who, in 1189, founded a "Brotherhood of the Bridges" whose self-imposed duty it was to maintain bridges without monetary reward.

The revival of vaulted stone bridge construction during the twelfth century was not entirely unrelated to the contemporary development of vaulting for ecclesiastical buildings. This connection, however, was somewhat superficial and hardly extended to technical and formal details. The ubiquitous pointed arch of the Gothic churches was applied to bridges only in exceptional cases, for instance at the Truyère Bridge near Entraygues (thirteenth century) or the Tharne Bridge at Montauban (1291–1335). The bearing structures of stone

---

[1] cf. Gagliardi, "Geschichte der Schweiz", Zürich, 1934, Vol. I, page 164, and the literature referred to on that page.

bridges generally consisted of barrel vaults ; but there was a great variety in details. The individual spans of the same bridge often differed considerably, and many diverse forms of soffit were in use, from the stilted semi-circular arch to oval curves and very flat segmental arches. For instance, the rise of the Ponte Vecchio at Florence (Fig. 17), which was built about 1335–1345 by *Taddeo Gaddi* or, according to other sources, by *Neri di Fioravante*, is only about one-fifth of the span. Whilst the great majority of Roman bridges consisted of a more or less regular sequence of similar, semi-circular arches carrying a roadway which was horizontal at least in its central part, the medieval bridges were usually more closely adapted to local circumstances. Even heavy gradients of the roadway were accepted if, owing to foundation difficulties, it appeared desirable to provide one large (mostly central) span and several smaller spans. Most of the medieval bridges therefore possess a more pronounced individuality than Roman bridges, and differ more from each other than the latter.

A strongly humped bridge is hardly objectionable if the traffic consists exclusively of pedestrians and beasts of burden, as is often the case in mountainous regions. Occasionally, one encounters a bridge such as the Serchio Bridge near Lucca, built during the fourteenth century (see Fig. 24) which, with its boldly ascending arch, prompts the question whether its design has, in fact, been governed by utilitarian considerations only, or whether æsthetic considerations have also played a part. Such conception is reminiscent of Dehio's[1] thesis in regard to medieval castles which, though utilitarian in character, were also designed to impress the imagination. "Not only were they to be strong in *fact* but also in *appearance*. That is why their overtowering and awe-inspiring effect was deliberately magnified."

The boldest bridge design of the Middle Ages, and one of the most outstanding engineering achievements of that time, was the great bridge over the Adda at Trezzo, in Northern Italy, which spanned the river in a single arch of 236 ft. length and nearly 70 ft. rise. This bridge, built for Bernabò Visconti in

[1] Dehio : "Geschichte der Deutschen Kunst", Vol. II, page 297.

1370–1377, served as an approach to the ducal castle and was fortified with turrets. Owing to its strategic importance, the bridge did not last more than half a century. It was destroyed by the deliberate weakening of one of the abutments during a siege of the castle in 1416. Today, nothing is left of the bridge except the two abutments with small overhanging remnants of the vault. This vault, which was about 7 ft. 5 in. thick and nearly 30 ft. wide, represented a great step forward, compared with the bridges of Antiquity. Its span was more than twice as great as that of the bridge over the Nera at Narni which is believed to have the longest span of any Roman bridge. It was not before the second half of the nineteenth century that similar spans were again attained ; they were excelled only after the advent of modern concrete and reinforced concrete bridges.

The aspect of the bridge of Trezzo was presumably similar to that of the Scaliger Bridge (Ponte del Castello Vecchio) at Verona (1354–1356) which, unfortunately, became a victim of World War II.[1] That bridge had solid spandrels and forked battlements (Fig. 25). The greatest of its three spans (160 ft.), and certain spans of a few French bridges (Vieille Brionde, 178 ft., built between 1340 and 1480, destroyed in 1822) were among those bridges bearing the closest resemblance to the Trezzo Bridge, also as regards the length of span.

The picture of medieval engineering may be rounded off by a reference to hydraulic works. One of the most grandiose examples in this respect was the irrigation and draining scheme which was carried out by the free Italian communities in the Lombardic Plain. The construction of the "Naviglio Grande" of Milan was begun in 1179 ; it was resumed, after a long interruption, in 1257 and completed a few years later. An extensive system of canals protected, and still protects, the Lombardy Plain against flooding and served to distribute the clear water of the Ticino for irrigation purposes.

[1] Together with other bridges of Verona, including a Roman bridge, the Ponte del Castello Vecchio was destroyed by German troops during the night of the 25th — 26th of April, 1945, two days before the Italian armistice was concluded (military necessity ?). The bridge has since been restored to an exact replica of its original form, as a result of a laborious effort of retrieving the original material from the river bed.

3. THEORETICAL MECHANICS AND STATICS IN THE SCHOLASTIC AGE

The philosophy of scholasticism, which aimed at a conciliation between faith and religion on the one hand, and reason and logical thought on the other hand, promoted unwittingly the development, not only of mathematics, but also of exact science and mechanics, owing to the formal training of reasoned and logical investigation. In addition, scholastic philosophers like Albertus Magnus or Raimundus Lullus also concerned themselves with applied natural science. The last-named author's "Ars magna" was a "phantastic textbook on inventiveness and notional combinations" which aimed at a "mutual penetration and reorientation of all knowledge", at a "consummation of all intellectual and natural sciences" (Feldhaus).

Of greater importance to the development of natural science was that great adversary of scholasticism, *Roger Bacon* (1214– 1294), who had received his training at the two foremost places of learning of his time, the Universities of Paris (founded about 1200) and Oxford (founded about 1214). To him, the "Doctor mirabilis", Experience, Experiment and Mathematics are, as in modern times, the three pillars of natural science. Arithmetic, geometry, astronomy and experimental science were his subjects. His treatise, "On the Secret Forces of Art and Nature" (1242), contains, *inter alia*, the earliest description of gunpowder.

Gothic architecture, with its refined system of vaults and buttresses, has been called "petrified scholasticism". In reality, however, the learned doctors and speculating philosophers lived in a world of their own, which was probably rather remote from that of the skilled masons and ingenious builders. The practitioners of the building sites and guilds, whose superb statical feeling is borne out by the Gothic vaults, must have been instinctively familiar with the laws of the lever, of the inclined plane, etc., long before the learned professors and mathematicians endeavoured to find rational, theoretical explanations and proofs for them. In their attempts to seek the truth, the scientists were sometimes induced by paralogisms to put forward assertions which the technical feeling of the practical craftsman was immediately bound to reject as false.

However, for the first time after a millenium of complete silence, the scientists of that age again discussed those problems which were to form the nucleus of an uninterrupted development — from then on — of the actual science of civil engineering — the "structural analysis".

At the dawn of medieval statics, one particular name is mentioned again and again : that of *Jordanus de Nemore* or *Jordanus Nemorarius*. Nothing is known about his person ; neither the times of his birth and death nor his nationality can be stated with certainty. He is said to have lived during the eleventh or twelfth century, but most historians assign him to the thirteenth century. It appears that he hailed from Germany and belonged to the Dominican Order. But, as Duhem points out, his cognomen "de Nemore" may also signify Italian nativity.

Not all of the many existing manuscripts on arithmetic, geometry and statics, ascribed to him, have been written by Jordanus himself. It is likely that the famous name has later been bestowed on treatises of unknown authors.[1] The work on problems of statics which must occupy us, for a moment, in connection with the subject of this book, has come down to us in different versions, and even under different titles : "Elementa Jordani super demonstrationem ponderis," "Elementa Jordani de ponderibus", and others. In this work, the author goes back to the few then known remnants of Greek learning, especially to the fragmentary treatise "De ponderoso et levi" which is ascribed to Euclid, and to Aristotle's "Mechanical Problems" (cf. page 21). In several respects, however, Jordanus was considerably ahead of his Greek models.

At the beginning of his treatise[2], he develops the concept of "gravitas secundum situm", that is to say, the gravity dependent on the position of the object. By this he means, in contemplating a weight movable on an inclined track, the one and only active component of the weight parallel to the track. The flatter the inclination, the smaller is the "gravitas secundum situm". This notion leads to reflections on the rectilinear and angular lever. These reflections, however, include certain fallacies which prove that the notion of the static moment was still unknown to him.

[1] cf. Duhem, Vol. I, page 98.
[2] cf. Duhem, Vol. I, Chapter VI.

On the other hand, Jordanus supplies a proof for the principle of the rectilinear lever with unequal arms which implies the intuitively assumed, though not expressly formulated proposition : "A force able to lift a weight G by the height h is also able to lift a weight G/n by the height n × h." This proposition contains the germ of the Principle of Virtual Displacements which was to play an important part later on, during the seventeenth and eighteenth centuries.

A great step forward, beyond the treatise just referred to, was another important work of the thirteenth century which has come down to us under the title : "Liber Jordani de ratione ponderis", but which, according to Duhem,[1] must have been written by a geometrician other than Jordanus. Because of the profound influence of this work on posterity, and particularly on the mechanical notions of Leonardo da Vinci, Duhem calls the unknown author "Leonardo's Precursor".

In the first three chapters, the author deals with problems of statics. In doing so, he corrects Jordanus' error regarding the angular lever. The notion of "gravitas secundum situm" is more clearly defined, and transferred to the problem of the Inclined Plane which here, for the first time, is solved accurately. In his argumentation, the author makes skilful use of the Principle of Energy which Jordanus had already applied to a special case. The last chapter of the treatise is concerned with problems of dynamics, with special reference to the influence of the medium (water, air) on the motion of a body.

At the conclusion of the Middle Ages, yet another scientist must be mentioned : *Blasius of Parma*, a physicist who concerned himself with theoretical mechanics, with the problems of the lever and the inclined plane, and also with hydrostatics. About him, slightly more is known. He was a physician by profession, and taught astrology and philosophy at Bologna, Padua and Pavia during the last quarter of the fourteenth century, and the first decade of the fifteenth century. His "Tractatus de ponderibus" has exercised an influence on the mathematicians of the Renaissance, Luca Pacioli, Leonardo da Vinci and Cardano.

[1] Volume I, Chapter VII.

CHAPTER III

THE BASIC PROBLEMS OF STATICS AND THE BEGINNINGS OF THE THEORY OF THE STRENGTH OF MATERIALS

1. LEONARDO DA VINCI AND THE MATHEMATICIANS OF THE RENAISSANCE — SIMON STEVIN

Among the multitude of physical and mechanical problems engaging the scientists of the sixteenth and seventeenth centuries, there are. two which are of fundamental importance to civil engineering : the *composition of forces* (force parallelogram) which is the basic problem of statics, and the problem of *bending* which is the most important task of the elementary theory of the strength of materials. Both problems were first tackled during the Renaissance, not by engineers or others directly concerned with building, but, as already pointed out in the introduction, by scientists as part of the general evolution of the exact natural sciences, and of the advent of mechanics as a mathematical, if empiric, discipline. The problem of force components was tackled by Leonardo da Vinci, Stevin, Galilei, Roberval and others before it was finally solved by Varignon and Newton. The problem of bending is also linked with some of these names and, in addition, with those of Mariotte, Hooke, Jakob Bernoulli, Leibniz, Parent and others ; but none of them succeeded in finding the final solution which had to await the advent of the eighteenth century (Coulomb and Navier).

For the successful solution of the basic problem of statics, the *composition of forces*, it was necessary, first of all, to conceive and define the notion of "force" as a directional ("vectorial") unit. This, in itself, was a task which called for a considerable effort of abstraction. At the outset, little distinction was made between "force" generally and weight ("gravitas") as a special form of "force". The geometricians and physicists of scholastic-· ism, Jordanus de Nemore and his successors, who were the first, since Antiquity, to concern themselves with problems of statics, had still but a vague notion of non-vertical forces ("gravitas secundum situm").

At the dawn of the Renaissance, a number of architects and plastic or graphic artists began to direct their attention to physical and mechanical problems. For a short while, it seemed as if some contact would be made between exact science, mathematics, geometry and mechanics on the one hand, and practical building construction and industrial technique on the other hand. The enthusiasm for Antiquity and the acquaintance with ancient authors on technical and scientific subjects tended to widen the intellectual horizon of the practising artists in the field of theory. The study of the laws of perspective and their subsequent application to painting called for a deeper knowledge of geometry. To this was added, as a general phenomenon of the Renaissance, a passionate craving for universal comprehension and knowledge, and an aversion to narrow specialization.

Leon Battista Alberti (1404–1472) endeavoured to reconcile theory and practice, science and life.[1] In his "Ludi matematici", written for the requirements of the practical technician, he offers guidance for the solution of concrete tasks in the sphere of surveying and engineering, such as the measurement of heights, the calculation of the width of a river by means of observations from one bank, the regulation of rivers, and the like.

The artist of the Renaissance most intimately concerned with theoretical mechanics and statics was *Leonardo da Vinci* (1452–1519). In his notes, we find not only sketches of innumerable machines and mechanical devices but also geometrical figures to illustrate theoretical relationships, or to derive and explain physical laws. He deals with the centre of gravity, the principle

[1] cf. Olschki, Volume I, page 45.

of the inclined plane, the essence of *"force"*.[1] The contemplation of a weight suspended on two inclined strings leads him to the notion of a force of optional direction, concentrated along a straight line, and to the problem of composing two forces to a single resultant force (Fig. 19). In this connection Leonardo

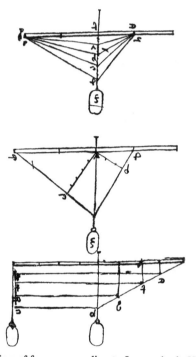

Fig. 19. Composition of forces, according to Leonardo da Vinci.
From *"Cod. Arundel", fol.* 1 *verso.*

applies, for the first time, the fundamental conception of the *statical moment* (force, multiplied by the distance from the fulcrum) to *non-vertical* forces. (For vertical loads, the notion

[1] Leonardo's definition of "force" is typical of the artist's picturesque language and dramatic impressiveness : "Forza dico essere una virtù spirituale, una potenza invisibile, la quale per accidentale esterna violenza è causata dal moto e collocata e infusa nei corpi, i quali sono dal naturale uso retratti e piegati dando a quelli vita attiva di maravigliosa potenza. . . ." (Cod. A, 36, V, quoted by Marcolongo, page 307).

of the statical moment is implicitly contained in the principles of the lever as defined by Archimedes and, much later, by Jordanus de Nemore and his successors). Another of Leonardo's sketches shows that he also attempted to account for the problem of the bending of a beam.

To what extent can Leonardo be regarded as an engineer ? The great painter, in common with all plastic and graphic artists of his age, sought close contact with architecture. We know of church models and town planning sketches designed by him, and also of fortification and canalization works which he planned and even, to some extent, actually supervised. But he was hardly active as a practising architect, in any case to a lesser degree than other contemporary Italian painters and sculptors, notably Michelangelo and Raphael. No buildings, or parts of buildings, can with certainty be ascribed to him. His architectural imagination manifested itself exclusively in his drawings and sketches.

Leonardo is the perfect prototype of the "uomo universale", the versatile man of the Renaissance. His craving for knowledge and cognition comprehended all branches of the exact and descriptive sciences. But his research work was non-systematic ; it was called forth by chance impulses from outside. As an artist and visualist, he attached great importance to immediate perspicuity, and in this respect he was indeed more akin to the technician and designer than to the exact scientist. Science was to be the servant of technique. However, in the case of most of the machines of his invention, he was content with a cursory graphic presentation, a hastily drawn sketch, without giving a thought to its actual practical application and efficient utilization. This fact tends to show that, as L. H. Heydenreich[1] has pointed out, Leonardo was mainly anxious to prove the utility of a theoretical conception by showing the *possibility* of its practical application. Once this proof was supplied, he turned to other problems. Intrinsically, he was less of a technician and inventor than it may appear. He was, first and foremost, concerned with *cognition*, and in spite of the constant emphasis on technical utility, his research was more

[1] Lecture held at the German Institute for the History of Art, Rome, on 19th May, 1941.

determined by principles than by purpose. "Naturalmente li omini boni desiderano sapere", even without being driven to it by practical necessity or external needs. If real practical tasks have inspired him at all, they are most likely to be found in the field of building construction such as the lifting and transport of heavy weights and the like. These tasks may have induced him to a closer study of the fundamental principles of mechanical devices and elementary machines like the lever and tackle, the influence of friction and similar questions.

Leonardo's many and, sometimes, detailed drawings of machine tools, ranging from metal working machines, lathes and draw benches, etc. to textile machines, mechanical spinning wheels, cloth-cutting machines and the like, do indeed point to his close acquaintance with all branches of contemporary technology. On the other hand, in spite of the title "Ingegnere generale" which he acquired in the service of Cesare Borgia, the predominance of his artistic temperament over his engineering mentality is evident from his obvious aversion to concentrating on any one given task, or working to a time schedule (also as a painter or sculptor), or even completing a work once begun. A true engineer is distinguished not only by a creative imagination but just as much by systematic thought and by a clear conception of technical and economic practicabilities. Of these traits, Leonardo possessed only the first, but that in rich abundance.

Leonardo was acquainted with Euclid's "Elements", as well as Aristotle's "Mechanics" and several of Archimedes' and Hero's writings. But his main source on statics was the treatise, "Liber Jordani de ratione ponderis" (cf. page 53) and the work of Blasius of Parma. Leonardo may owe the acquaintance with these sources to the mathematician, *Luca Pacioli* (1445– about 1515), a friend of his during his stay at the Court of Lodovico Moro in Milan. In one of his notebooks we find an entry to the effect that he acquired Pacioli's "Summa de Arithmetica" for the amount of 119 soldi.

As is well known, the great painter did not publish any scientific work during his lifetime. His ideas, disquisitions and discernments, some of which anticipated the results of much later research, were buried in his diaries. It was only at a much later date that these notes were published and thus made accessible to a wider circle of readers. They are, in a way, soliloquies which hardly exerted any immediate influence on the further development of science. If such influence existed, it is difficult to trace. Duhem, it is true, believes to have detected a channel through which Leonardo's ideas may, in fact, have served as a direct inspiration to his successors. He advances a thesis which, though not proven, appears to be well substantiated. According to this thesis, copies of the manuscripts which the master had bequeathed to his pupil, Francesco Melzi, and which were kept in the latter's villa at Vaprio near Milan, came to the knowledge of Cardano, Benedetti, Guidobaldo del Monte and Bernardino Baldi and have thus exerted an indirect influence on Galilei.[1]

During the first part of the sixteenth century, architecture and mechanics, treated as one by the "searching masters" of the early Renaissance, part company once more. To Michelangelo as well as to the classicists, Vignola and Palladio, architecture is again predominantly a matter of æsthetics. At the same time, mechanics and statics become a prerogative of specialist scientists. In the Italy of the sixteenth century, the occupation with Antiquity, particularly the study of Greek, also resulted in an advance of the exact sciences. Euclid, Archimedes, Hero, Apollonius, Ptolemy, Pappus, Aristarchus were translated into Latin and annotated, for instance, by *Federico Commandino* (1509–1575) of Urbino, teacher of a generation of "humanistic geometricians", who was just as famous for his knowledge of Greek as for his knowledge of the mathematical sciences. His pupil, *Bernardino Baldi* (1553–1617), translated and edited works of Hero, and wrote commentaries to the works of Vitruvius and Aristotle's "Mechanics".

Most of these mathematicians and geometricians of the Italian Renaissance have concerned themselves, *inter alia*, with problems of mechanics. Among them were *Cardano* (1501–1576), *Tartaglia* (sixteenth century), *Benedetti* (1530–1590), *Bernardino Baldi* and another pupil of Commandino's, *Guidobaldo del Monte* (1545–1607). In the sphere of statics, they

---

[1] cf. Duhem, Vol. I, page 35 ; Vol. II, pages 110 and 139 ; also Mach, page 76.

studied, like Leonardo, the theories of the lever, the inclined plane, the pulley, etc. They attempted to find a mathematical formulation for these laws and to arrive at a scientific, theoretical understanding of the working of such simple machines as the windlass, screw, pulley tackle, etc. Descriptions and theoretical explanations of these machines are contained in Guidobaldo's "Mechanicorum Liber" (Pesaro, 1577) and Benedetti's "Diversarum speculationum mathematicarum et physicarum liber" (Turin, 1585).

Among all these scientists, it was Tartaglia who was most concerned with practical aims. His "Quesiti et inventioni diverse" deal, among other things, with problems of artillery and ballistics, fortifications and the like. The work is a kind of collection of recipes for artists and technicians. Tartaglia and Cardano (his opponent in a priority claim for the discovery of the solution of cubic equations) are more in line with Jordanus de Nemore and his school, and are ahead of the peripatetic school. Benedetti and Guidobaldo, on the other hand, go back to the Greeks, to Aristotle, Pappus, etc., and cherish the subtle argumentations of Euclid.

Those interested in further particulars of this early stage of statics are referred to P. Duhem's excellent work, "Les origines de la statique" (Paris, 1905–1906), where the contributions of each of these scientists are described and analyzed in detail.

The scientific ideas of the Italian Renaissance soon exerted their influence beyond the confines of that country. They came, *inter alia*, to the knowledge of the Dutchman *Simon Stevin* (1548–1620) who, in his writings, quotes and criticizes, amongst others, Cardano's "Opus novum de Proportionibus". In addition, Stevin uses the same ancient and medieval sources as Leonardo and concerns himself with the same static problems as his Italian contemporaries — centre of gravity, lever, inclined plane. He, too, discusses the composition and resolution of forces in the case of a weight suspended on two strings. In his work on statics which, in 1586, was published in Dutch and, in 1608, in a Latin translation under the title "Mathematicorum Hypomnemata de Statica", he makes correct use of the parallelogram of forces and of the notion of the static moment.

He was the first to represent the magnitude of a force graphically as the length of a straight line parallel to the direction of the force. In doing so, he laid the foundation of "Graphic Statics", that branch of engineering science which, during the nineteenth century, was to become an important part of structural analysis.

The special importance of the Dutch mathematician lies in the strength of his imaginative power which leads him, through mere intellectual abstraction, with instinctive certainty to logical conclusions. A case in point is his grasp of the problem of the inclined plane. He imagines an endless chain laid around a three-edged prism (see Fig. 20), arguing that this chain

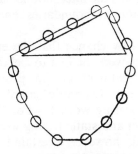

Fig. 20. Inclined plane with endless chain, according to Stevin.

would obviously retain its equilibrium even in the absence of friction ; this reasoning leads him logically to the law of the inclined plane.[1]

In the field of hydrostatics, Stevin is the first to pronounce the so-called "hydrostatic paradox", to the effect that a liquid may in certain circumstances (viz. if the containing vessel is smaller at the top) exert a total pressure on the bottom of the vessel which exceeds the aggregate weight of all the liquid in the vessel.

Stevin had commenced his career as a merchant, first at Antwerp, later with the administration of the Free Port of Bruges. In later years, after extensive travels in Prussia, Poland, Sweden and Norway and after studies at the University of Leyden (1583), he became a Professor of Mathematics at The

[1] cf. Mach, page 24 and Duhem, Vol. I, page 272.

Hague and a Treasurer in the service of Prince Moritz of Orange. His further duties included those of a Quartermaster of the Army, Inspector of Dikes and Superintendent of Waterways. In this capacity he had occasion to meet practical engineers in the sphere of earthworks and waterways construction. On the whole, he was more inclined towards practical aims and tried to make his scientific efforts serve the needs of everyday life.

Apart from his books on problems of mechanics, Stevin was also the author of works on book-keeping, military matters and other subjects, all in Dutch. The most important of them were soon translated into Latin, which at that time, and for a long time to come, was the international language of scientists, not confined to national boundaries and not restricted by national prejudices.

It is perhaps not by chance that the engineers engaged in hydraulic engineering were the first to feel the need for a mathematical, scientific treatment of the problems of mechanics and statics. (Like Stevin, Leonardo had also concerned himself with hydraulic engineering works). For the design of domes and bridges, intuition and static feeling were sufficient. For the construction of canals, however, a certain knowledge of hydraulics as well as accurate field surveys were indispensable, and these, though based on primitive instruments, could only be carried out by mathematically trained surveyors.

2. GALILEO GALILEI

With some justification, the majority of the Italians of the Renaissance referred to above could be called "ingenious dilettantes". A short time later, however, Italy produced one of the greatest scientists of all times in the sphere of mechanics, *Galileo Galilei* (1564–1642 ; i.e. a slightly younger contemporary of Simon Stevin). His fundamental achievement is his reliance on direct observation and experiment rather than on the conventional, blind faith in the authority of the ancient scientists, especially Aristotle. He, it may be said, substituted the question "How ? " for the question "Why ? " The Tuscan physicist's achievements are of vital importance not only to dynamics, astronomy and optics, but also the scientific branches of civil engineering proper, namely, statics and the theory of the strength of materials. It is therefore befitting to devote some space to him.

In his youth, Galilei followed the usual humanistic course of studies. In addition to Plato and Aristotle, the then known works of Archimedes and their methodical combination of experiment and mathematics attracted him greatly. Apart from that he owed his mathematical instruction, and the foundation of his ideas on mechanics, to the Italian mathematicians of the sixteenth century. In his writings, he quotes Commandino and Guidobaldo (in the "Discorsi e dimostrazioni matematiche intorno a due nuove scienze", end of the fourth day) as well as Cardano's "De subtilitate". No doubt he also knew the latter's "Opus novum".

Having studied medicine, philosophy and mathematics at the University of his home town, Pisa, Galilei was appointed a Professor of Mathematics at that University at the early age of 25 years. Even then, during his three years' teaching at Pisa, he occupied himself extensively with problems of mechanics, especially the law of falling bodies and the law of the pendulum. In 1590, he proved experimentally the fallacy of Aristotle's proposition : "Bodies of different weight fall at different speeds."

In 1592, Galilei moved to the University of Padua where he obtained a better paid professorship, likewise for mathematics which, at that time, included mechanics and astronomy.

During the time of his activities at Padua, Galilei became gradually convinced of the truth of the new "world system", propounded by Copernicus. Whilst he still subscribed to the traditional Ptolemaic System in his lectures, as is shown by some copies still in existence, he privately professed the new doctrine. With the aid of the telescope perfected by him, he made, during the last years of his stay in Padua, the revolutionary astronomic discoveries summarized in his work "Sidereus Nuncius" which was published in 1610 and dedicated to the Grand Duke Cosimo II of Toscana. This dedication resulted in a summons to the court of Florence where he became a "First Mathematician" at an annual salary of 1000 Tuscan thalers.

Well known are Galilei's later adventures in Rome where

he was summoned in 1633, at an advanced age, to appear before the Inquisition Tribunal because of his endorsement of Copernicus' doctrine and the publication of his "Dialoghi sui massimi sistemi". These events are of interest in connection with our subject in as far as they may have been the indirect cause of Galilei's greater interest in mechanical problems to which he devoted his last work. With the exception of the posthumously published "Mechanics", all his earlier works were mainly concerned with astronomic questions. In consequence of the Roman verdict, and the constant supervision on the part of the Inquisition, he was no longer allowed to express his opinion, "in writing or speech or in any other way, on the movement of the earth and the immobility of the sun". The aged scientist therefore reverted, of necessity, to the less dangerous problems of mechanics and statics with which he was concerned during his lectureship at Pisa and Padua. The results of his research and experience in this sphere are once more summarized in his "Discorsi e dimostrazioni matematiche intorno a due nuove scienze". The publication of this book was, incidentally, prohibited in all Catholic countries ; it was published by Elzevir in Leyden in 1638.

In the sphere of *statics*, our main debt to Galilei is a clearer notion of "*force*" and "*statical moment*". The latter conception goes beyond that of Archimedes in as much as it is applied to forces pointing in any direction. Galilei was, moreover, the first to use the term "moment",[1] though still in the wider sense of the word, meaning the "effect of a force" generally speaking. He applies the term to the product of a weight and the speed with which it moves, as well as (in connection with statical problems such as the lever or the balance) to the statical moment proper, i.e., the product of a force and its distance from the fulcrum. In either case, the same weight is capable of exerting a greater or smaller effect, all according to the speed in the one case, and to the distance from the fulcrum in the other case.

---

[1] Olschki (II, page 74) stresses the significance which must be ascribed to the coining of a new technical term : "The very act of coining a new term shows that the scientist and thinker has become conscious of the originality of his thoughts ; he becomes a real innovator only when he has enriched science by a new notion, and the language by a new word, or a new meaning of a known word."

As already mentioned, Galilei's contributions to our knowledge of statics are based on the work of his predecessors, such as Archimedes, Jordanus de Nemore, Leonardo da Vinci, Cardano and others, whose doctrines were revised by him, and more clearly formulated. His share in the development of dynamics, especially the formulation of the laws of falling bodies, may here be ignored as being outside our special subject. In another field of mechanics, however, he covered entirely new ground. If we except Leonardo's tentative notes on this subject (see page 57), Galilei was the first to discuss the *bending strength* of a beam. He thus became the founder of an entirely new branch of science : the *theory of the strength of*

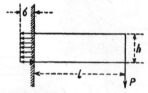

Fig. 21. Flexure, according to Galilei.

Reproduced from *Schweizerische Bauzeitung*, Vol. 116.

materials, which was destined to play a vital part in modern engineering science.

Galilei starts with the observation of a cantilever beam, subjected to a load at the free end. To this beam, he applies the principle of the angular lever (Figs. 21 and 22). He equates the statical moments of the external load with that of the resultant of the tensile stresses in the beam (which he assumes to be uniformly spread over the entire cross-section of the beam) in relation to the axis of rotation (assumed to be located at the lower edge of the embedded cross-section). He thus arrives at the correct conclusion that the bending strength of a rectangular beam is directly proportional to its width, but proportional to the square of its height. But, as Galilei bases his proposition merely on considerations of statics and does not yet introduce the notion of elasticity, propounded by Hooke half a century later, he errs in the evaluation of the *magnitude* of the bending strength in relation to the tensile strength of the beam.

Expressed in modern terms, the moment of resistance of the rectangular beam would, according to Galilei, amount to $\frac{bh^2}{2}$, which is three times as great as the correct value, $\frac{bh^2}{6}$.

Galilei's contribution to the theory of the strength of materials consists mainly in the inspiration which he imparted

Fig. 22. Cantilever beam, loaded at the free end. From Galilei, *Discorsi e dimostrazioni matematiche*, Leyden, 1638.

Reproduced from *Schweizerische Bauzeitung*, Vol. 119.

to his successors. His basic achievement lies in the correct, original formulation of the problem which was, under the name of "Galilei's Problem", to engage the minds of scientists through two centuries, until Coulomb and Navier succeeded in finding the correct answer.

By way of postscript to our contemplation of Galilei, we may point to the fact that his works, written in an original and lively style, have come to play an important part in Italian literature and have thus found their way to a wide circle of readers, not only for the sake of their importance to the history of science but also on account of their literary and artistic merits. His "Discorsi" is thus perhaps the only book within the sphere of engineering belonging to that small number of immortal works which, beyond their immediate impact on contemporary science, have become the intellectual property of the cultural élite of a whole people. In this respect it may, for instance, be likened to the works of Machiavelli or Jakob Burckhardt in the field of historical science, or those of Alexander von Humboldt in the sphere of geography.

3. FRENCH AND ENGLISH SCIENTISTS OF THE SEVENTEENTH CENTURY

In the development of exact science, including the branches of engineering science of special interest to us, as well as in other spheres of cultural evolution, we notice, from the beginning of the seventeenth century onwards, a marked shift from Italy to the countries North of the Alps, especially France and England. The interruption, caused by the Thirty Years' War, of the Renaissance-inspired intellectual evolution in Germany, is evident from the fact that the names of German scientists are almost completely absent from the history of theoretical mechanics and its application to civil engineering, up to the end of the eighteenth century.

In the sphere of statics, the principle of the parallelogram of forces was followed up again, half a century after Stevin, by *Roberval* (1602–1675). In his "Traité de méchanique", which was later embodied in Mersenne's "Harmonie universelle", the Frenchman re-states the principle and tries to prove it, somewhat diffusively, with the aid of the notions of the lever and the inclined plane. Towards the close of the seventeenth century, Lamy, Varignon and Newton, approaching the problem from the viewpoint of dynamics, arrive at the formulation which is still familiar to us today. *Newton*, in particular (cf. page 75), applies the proposition mainly to problems of dynamics, using it to explain the motions of celestial bodies in the terms of his general law of gravitation. His classical formulation which, on the Continent also, is the one most widely used, is summed up by Mach in the following sentence : "If a body is simultaneously subjected to two forces, which would, in a given time, cause the motion A — B and A — C, respectively, the body will during that time move from A to D" (Fig. 23).

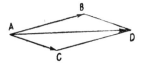

Fig. 23. Parallelogram of forces, according to Newton.

Varignon (1654–1722), however, still follows Aristotle's assumption that a force is proportionate to the speed. As the Parallelogram of Velocities was already well known, he has no difficulty in applying the same principle to forces, and thus to statics at large[1].

But Varignon owes his fame mainly to the fact that he was the first to pronounce the doctrine of the superposition of statical moments : "The moment of the resultant of two forces, related to any given point, is equal to the algebraic sum of the moments of the two component forces." This fact was known before his time, but Varignon was the first to generalize and formulate it.

Another elementary principle of statics, known at a fairly early stage, was that of the *virtual displacements*. Already in the Middle Ages, Jordanus de Nemore and his school (cf. Chapter II, Section 3) and afterwards, the mathematicians of the Renaissance (Guidobaldo del Monte) had applied this principle to theoretical considerations, departing from its most obvious manifestation on the lever and tackle. To apply the postulate of the equality of motion-causing and motion-resisting energies to statical problems (e.g. in order to prove the propositions regarding the composition of forces or statical moments) is, in fact, nothing but an anticipation of the methods of virtual displacements. Stevin and Galilei used the principle to explain the working of elementary machines. The latter, and after him mainly *Descartes* (1596–1650), already use in their treatises

---

[1] Duhem, Vol. II, page 255.

the notion, though not yet the term, of "energy". Galilei talks of "Momento", whereas the French philosopher applies the term "force" to the product of "weight" and distance covered. Descartes then takes the important step of regarding, not *any* motions, but *infinitely small* motions, insisting expressly that it is the *beginning* of a motion that must be considered. "La pesenteur relative de chaque cors se doit mesurer par le commencement du mouvement que devrait faire la puissance qui le soutient, tant pour le hausser que pour le suivre s'il s'abaissait."[1]

The final formulation of the Principle of Virtual Displacements appears in a letter which *Johann Bernoulli* (1667–1748) wrote to Varignon on 26th January, 1717, and which the latter embodied in his "Nouvelle Mécanique" : "Concevez plusieurs forces différentes qui agissent suivant différentes tendances ou directions pour tenir en équilibre un point, une ligne, une surface, ou un corps ; concevez aussi que l'on imprime à tout le système de ces forces un petit mouvement . . . chacune de ces forces avancera ou reculera dans sa direction, . . . ce que j'appelle vîtesses virtuelles. . . . En tout équilibre de forces quelconques, . . . la somme des Energies affirmatives sera égale à la somme des Energies négatives prises affirmativement."[2]

As regards the problem of *bending*, the greatest step forward, after Galilei, was made by the English physicist, *Hooke* (1635–1703), who observed that the force with which a "spring" attempts to regain its natural position is proportional to the distance by which it has been displaced. Hooke emphasizes expressly that this proposition applies not only to coiled springs but to any "elastic body", to metals, wood, stone, silk, bones, glass, etc., and to tensile as well as compressive stresses.[3] Hooke's proposition "Ut tensio sic vis" means nothing else but the Principle, still named after him, of the proportionality of the strain produced and the stress producing it. It is on this principle that the further development of the classical theory of strength and elasticity of materials is based.

Mariotte (1620–1684) reverts to Galilei's Problem and,

[1] Duhem, Vol. I, page 337.
[2] Duhem, Vol. II, page 267.
[3] Hooke "De potentia restitutiva". London, 1678, cf. Todhunter, Chapter I.

applying Hooke's new method of approach to the fibres of a beam subjected to bending, conceives the idea of assuming a triangular distribution of the internal stresses. To begin with, he still assumes, with Galilei, that the bending moments are related to a point at the lower edge of the embedded cross-section. Later on, he corrects this error and places the "axe d'équilibre"[1] correctly into the middle of the rectangular cross-section. As Saint-Venant points out, the basic principle of the calculation of bending was herewith established, although Mariotte's further arithmetical calculation was marred by an error.

Saint-Venant and Todhunter[2] have described how, after Mariotte, other scientists such as Jakob Bernoulli, Leibniz and Varignon concerned themselves with this problem, without coming much nearer to the correct solution. Bernoulli makes the same mistake as Mariotte and concludes that the position of the neutral axis is indifferent. Leibniz and Varignon revert, in this respect, to Galilei, placing the neutral axis at the lower edge of the cross-section. They do, however, assume with Mariotte, that the stretched fibres are elastic, and that the distribution of the tensile stresses is triangular. The idea of taking the elasticity of the fibres into consideration was therefore called, by contemporary writers, the "Theory of Mariotte and Leibniz".

Parent (1666–1716) is repeatedly concerned with the problem, again assuming, like his predecessors, that the axis of rotation lies at the lower edge of the cross-section. Only in his last treatise, written in 1713, does he state "that a single line at the lower edge of the cross-section cannot offer sufficient resistance to serve as a support during the rotation, as the sum total of the pressure resistance at the cross-section must be equal to that of the tension resistance". This discovery permits him to correct the error committed by Mariotte and Bernoulli and to find, for the rectangular beam, the correct relation of bending strength and tensile strength.[3]

---

[1] The term "neutral axis" is first found with Tredgold (1788-1829) ; cf. Saint-Venant, para. 4.
[2] See Bibliography at the end.
[3] cf. Saint-Venant, para. 6.

If we cast our mind back to the men who have created the basis of statics and of the theory of the strength of materials from the onset of the sixteenth century to the middle of the eighteenth century, we notice that there is not a single architect or bridge builder among them. True, the architects and artists of the early Renaissance often displayed a vivid interest in geometry and mathematics, as we shall see in the following chapter. As regards the actual creators of the *science* of *statics*, most of them were still physicists, mathematicians and geometricians all in one.[1] But we have absolutely no evidence that any of them has played a decisive part in the erection of major buildings. Structures such as the dome of S. Peter's in Rome, $137\frac{1}{2}$ ft. in diameter (1588–1590), the castles and churches of the two Mansarts, the 66 ft. barrel vault of S. Michael, Munich (1583–1597) — the examples could be multiplied *ad libitum* — would nowadays be regarded as tasks to be undertaken by a civil engineer. They were, in fact, erected by master builders who had but little contact with the contemporary theoreticians of mechanics.[2] The *practical builders* were excellent structural designers with a highly developed sense for statics who were otherwise constrained to rely on a few empiric rules. These were summarized in numerical form by certain architectural theoreticians, such as Leon Battista Alberti (see Chapter IV, Section 2). True, the architect was always required to possess a certain knowledge of mathematics and

[1] Many of the Italian Renaissance scientists in the sphere of geometry and mechanics had been led to the natural sciences through the study of medicine. Blasius of Parma, Commandino, Cardano, Bernardino Baldi and Galilei had thus studied medicine in their youth, and one or two of them had, in fact, practised the medical profession. In Germany, it was a physician (Agricola, alias Georg Bauer, 1494-1555) who, in his twelve books "De re metallica", created the most comprehensive mining compendium of the Renaissance.

[2] An exception is the great English architect, Christopher Wren (1632-1723), who, appointed a Professor of Mathematics at Oxford at the age of 28 years, was actively engaged in research on mechanics, hydraulics and astronomy before he became a "Royal Architect", in 1668. In this capacity, he displayed an extremely fertile activity, being responsible for the design of about fifty churches, including S. Paul's Cathedral in London, as well as numerous public buildings, hospitals, Royal palaces, etc. As a town planner, he designed, after London's Great Fire in 1666, a plan for the reconstruction of the City which did not materialize, however. Also Claude Perrault (1613-1688), the designer of the Louvre façade, published *inter alia*, a treatise on "Recueil des machines" (1700) and one on "Essais de physique" (1680-1688).

mechanics. This knowledge, however, was needed not for the structural analysis and the determination of dimensions and stresses, but rather for the mathematical rules of design composition, the survey and measuring of the building and possibly to facilitate the construction of transport and hoisting devices. This may explain a note by the translator of the French edition (1634) of Galilei's "Mechanics" who described his own additions and annotations as "useful to *architects*, well-sinkers, philosophers and artisans".

The advance of the *science* of mathematics and mechanics, on the other hand, was the work of scientists, physicists and geometricians who either held teaching appointments at colleges or universities or received a salary from their sovereign, with or without the obligation to teach. In their capacity as "Court Mathematicians",[1] some of them may have had to deal with public engineering works, such as the construction of canals, harbours or fortifications, or with mining, which would have afforded them opportunities to make use of their knowledge of geometry and surveying. But it is not likely that their interest in problems of statics and of the strength of materials was promoted by their professional activities.

In France, in particular, the advance of mechanics was paralleled, and to some extent influenced, by the development of higher education in general. The "Académie des Sciences"[2] was founded by Colbert in 1666, and enabled numerous scientists to produce and publish their works. The "Collège de France", in existence since 1530, was joined by other colleges, such as "Collège Gervais", or the "Collège Royal".

Roberval was, since 1631, Professor of Mathematics at the Collège Gervais, later at the Collège Royal in Paris, and a Member of the Academy since its foundation.

Varignon, too, having been a keen student not only of theology for which he was destined, but also of philosophy and geometry, became a Professor of Mathematics at the Collège

---

[1] Baldi was Court Mathematician at Urbino, Benedetti at Turin, Galilei at Florence.

[2] The "Académie d'Architecture", founded by Colbert in 1671, was more concerned with the formal-artistic aspects of architectural training, and was of much lesser importance to the development of statics than the various Colleges at which mathematics and physics were taught.

Mazarin in Paris, and a Member of the "Académie des Sciences".

Mariotte had likewise taken orders early in life. He became a Prior of the Monastery of Saint Martin sous Beaume, near Dijon. In 1666, he became a Member of the Academy. Otherwise, little is known about his life.

Parent was a student of law before he turned to mathematics and physics, and particularly mechanics, the subject to which most of his publications are related.

Duhem[1] gives a vivid description of one particular aspect of scientific activities at that time. The task of disseminating information, nowadays performed by the technical Press, was at that time performed by correspondence between the scientists. During the first half of the seventeenth century, it was mainly the Franciscan, *Mersenne* (1588–1648), who conducted an extensive correspondence with nearly all contemporary geometricians and physicists of France, with the express purpose of collating and publishing as much of contemporary science as possible. It is thus in Mersenne's books that one has to look for the statical achievements of Roberval who published hardly anything himself.

At the dawn of Mechanics, the scientists, from Jordanus to Varignon, were mainly endeavouring to find strict and flawless proofs, in the Euclidean manner, for the elementary laws of statics (lever, inclined plane, parallelogram of forces, equation of moments, etc.). These often rather circumstantial argumentations and conclusions formed the principal contents of contemporary literature and scientific correspondence ; they are of but limited interest today and will therefore not be discussed here in greater detail. Mach has this to say on the subject : "Small wonder if the discoverer or tester of a new rule, driven by distrust against himself, is looking for a *proof* of the law which he believes to have discovered. . . . A new law may be proved by comparing it with frequent experience. . . . But the discoverer wishes to achieve his aim more quickly. He compares the result of his own rule with all the experience familiar to him and with all the old-established

---

[1] Vol. I, page 311 and Vol. II, page 187.

and proof-hardened rules, checking up if there is any disparity. . . . But if, after a time, a rule has been proved correct sufficiently often, science ought to acknowledge the fact that a further proof has become unnecessary, and that there is no point in regarding a rule as more safely established if it is supported by other rules which, in their turn, were established in the same empiric fashion, only earlier. . . ."[1] With this statement, Mach has in mind the principle of the static moments (principle of the lever) or the parallelogram of forces which are nowadays instinctively regarded as elementary principles whilst they were, at the time of their first formulation, carefully based on other propositions which, through a mere historical accident, were already familiar.[2]

If we talk, in this connection, about the influence of one scientist on another, this applies mainly to the manner and method of argumentations which were then used with great intellectual effort, to prove propositions which are nowadays taken for granted.

In this respect, Leonardo da Vinci may be regarded as "modern" in as much as, in contrast to his predecessors and immediate successors, he laid more emphasis on clear and intelligible formulation than on strict and irrefutable proofs. At a time when most of the conceptions now familiar to us, such as force, moment, component, etc., were neither clearly defined nor unequivocally termed, the formulation of a proposition was a much more difficult task than is imagined today.

4. THE MATHEMATICIANS OF THE RATIONALIST AGE

The great discoveries in the sphere of mathematics during the last third of the seventeenth century and the first half of the eighteenth century, especially the invention of the Infinitesimal Calculus, advanced and completed that intimate relationship of natural science and mathematics which has since

---

[1] Mach, pages 70-72.

[2] "Like Varignon, many scientists concerned with Mechanics wanted to derive the Principle of the Lever from the Parallelogram of Forces. Others, like Johann Bernoulli, were of the opinion that the Parallelogram of Forces should, conversely, be derived from the Principle of the Lever." (Rosenberg, "Geschichte der Physik", Vol. II, page 293.)

become the symbol and essence of the exact sciences, physics and mechanics.

The names of *Leibniz* (1646–1716) and *Newton* (1642–1727) have already claimed our attention (pages 70 and 67) in connection with the problem of bending and the problem of the composition of forces. The encyclopedic intellect of the German philosopher comprehended all fields of contemporary knowledge, ranging over mathematics, history, jurisprudence and theology. Not only was he, as a physicist, occasionally concerned with problems of mechanics ; through his connection with mining affairs at Brunswick, he also appears to have been in contact, for a time, with practical engineers concerned with underground works.

The English physicist has to his credit a wealth of achievements, including not only the formulation of the theory of gravitation and important discoveries in the field of optics and general physics, but also accomplishments which have an important bearing on civil engineering and structural engineering proper. Foremost among them are the definitions and propositions which concern the foundations of classical mechanics and which are elaborated in his main work "Philosophiæ naturalis principia mathematica" (London, 1687). Apart from the clear and general formulation of the principle of the Parallelogram of Forces (already referred to on page 67), Mach lists the following important achievements :

1. The generalization of the notion of "Force".
2. The formulation of the notion of "Mass".
3. The formulation of the Principle of Effect and Counter-effect.[1]

But the most important achievement of both men, also in regard to the development of mechanics and the theory of the strength of materials, was the invention of the *Infinitesimal Calculus*. As is well known, the foundation for the new calculus was laid during the last third of the seventeenth century, simultaneously but independently, by Leibniz and Newton. The priority controversy between the two great mathematicians caused excitement for many years, in the scientific world of

[1] Mach, page 186.

the early eighteenth century. Leibniz had approached the problem on the strength of philosophical logic and geometrical considerations, pondering over the notion of continuity and the conception of a mathematical function. Newton, the physicist, had arrived at the new calculus through his occupation with concrete problems of dynamics.

The new Differential and Integral Calculus represented a most elegant instrument of mathematics which was a temptation to seek further applications. The extraordinary versatility of the new calculus engaged the imagination, not only of experts, but of all educated circles during the Rationalist Age. "Nearly everybody regarded himself as an expert in mathematics. All the world had heard of the invention of the Infinitesimal Calculus, and of the controversy concerning the honour of its first discovery. And all the world was dazzled by the enormous success with which Newton had applied the new method to astronomy. As at the time of Plato, mathematics became popular. . . . In any case, mathematics became, during the eighteenth century, the principal branch of science which could not be ignored by any educated person even though he may have been inwardly averse to it, like Frederick the Great or Goethe."[1]

For a time, mathematics thus became not only the servant but also the inspiration of the exact sciences. The methods of the Infinitesimal Calculus were further developed, mainly by the Basle school of Leibniz's correspondents and admirers, namely the members of the Bernoulli family and Leonhard Euler. In their search for further applications, they found many interesting solutions to problems of mechanics and the strength of materials which, often much later, came to play an important part in practical engineering.

The most important case in point is that of the *elastic curve*. More than fifty years before Coulomb (cf. page 150) succeeded in finally solving the problem of bending, Jakob Bernoulli and Euler had already discovered and described the true nature of the "elastic curve". *Jakob Bernoulli* (1654–1704), the oldest member of the family, stated already in 1694, in contrast to

[1] O. Spiess, "Leonhard Euler", Frauenfeld, 1929, page 18.

earlier arbitrary and erroneous statements (e.g. by Galilei), that "the bending radius of an originally straight, homogeneous beam is, at any given point, inversely proportional to the moment of the bending force in relation to that point".[1]

A thorough investigation of the elastic curves was subsequently carried out by *Leonhard Euler* (1707–1783), the most significant and fertile mathematician of the Basle school of the eighteenth century. Having completed his studies at his home town, Euler spent the rest of his life at the newly established Academies of S. Petersburg (1727–1741), Berlin (1741–1766) and again S. Petersburg (1766–1783). In spite of his religious background, he was a typical son of the Rationalist Century, believing that, with the aid of mathematics, he could explain the whole world and ascribe its existence to mechanical processes. Numerous are the applications of his analysis to problems of physics, especially mechanics.

In 1744, Euler published an exhaustive treatise on plane elastic curves (bending lines), distinguishing between nine different cases for which he formulated differential equations. His solution of the problems is unobjectionable, except that he does not yet make use of the notions of the "modulus of elasticity"[2] and "moment of inertia", using, instead, the product $E \times I$, a single value depending on the properties of the beam. He named this value the "absolute elasticity" ("Elasticitas absoluta") and erroneously assumed it to be proportional to the square, instead of the third power, of the section height (in the case of rectangular cross-sections). One of the nine cases investigated yields, on integration, the famous "Euler Formula" which is still in use today to express the buckling strength of struts. The formula first appeared, in its usual form ($P_c = \dfrac{\pi^2 E I}{l^2}$, as it would be written today), in Euler's treatise "Sur la force des colonnes", published in 1757.

In his treatment of the elastic curves, Euler makes use of the so-called "Principle of Least Effort". Already before Euler's work appeared, *Daniel Bernoulli* (1700–1782) had stated in a letter to his friend and colleague of his Petersburg

[1] Saint-Venant, para. 12.
[2] cf. footnote on page 155.

period that the elastic curve of an originally straight beam, bent by external forces, will assume that shape at which the total bending energy becomes a minimum. This is the germ of the "Principle of Least Action" which was to acquire great importance in the methods of structural analysis developed during the nineteenth century (e.g. by Müller–Breslau, cf. page 221).

During the eighteenth century, it was a favourite method to solve all kinds of physical and mechanical problems by reducing them to the determination of a minimum or maximum ("Isoperimeter Problem"). The simple, quasi-automatic method of determining, with the aid of the Differential Calculus, the minimum or maximum value of an equation or function, is already elegant and impressive in itself. Euler, and later Lagrange, went further by developing the so-called "Calculus of Variations" which permits the determination, for given marginal conditions, of that function or curve for which a certain property assumes a maximum or minimum value. To an age largely theologically orientated, the occurrence of such curves in natural phenomena (e.g. the motion of light in a medium of variable density, or the elastic curve, etc.) appeared to be the realization of the metaphysical, theological principle of achieving the greatest effect with the least effort. This principle was deemed to be a particularly appropriate expression of the wisdom of Him who created "the best of all possible worlds."[1]

The *catenary curve*, too, can be regarded as a "minimum problem", in as much as a chain, suspended from two fixed points, will assume that form at which the centre of gravity is in the lowest possible position. The problem of determining the equation of this curve was put forward by Jakob Bernoulli in 1690 and was solved, a year later, by his brother Johann as well as by Leibniz and Huygens.[2]

At that time, the application of the "Principle of Least

---

[1] "It is . . . probable that physicists have to thank this theological tendency in great part for the discovery of the modern principles of Least Action, of Least Constraint and perhaps of the Conservation of Energy" (Todhunter).

[2] A few years later, the Scottish mathematician, D. Gregory, dealt with the catenary curve in a treatise entitled "Properties of the Catenaria", Phil. Trans., 1697 ; cf. page 140.

Effort" was regarded as an aim in itself, whilst the importance of the technical problems solved through this application was regarded as secondary. This is, for instance, apparent from the title of Euler's treatise, which was destined to play such an important part in the theory of the strength of materials. The famous buckling formula first appeared in an Appendix, headed "De curvis elasticis", to a work, published in 1744 and entitled "Methodus inveniendi lineas curvas *maxime minimeve proprietate* gaudentes". The primary aim was thus not the determination of the elastic curves, or the solution of the buckling problem and its practical applicability to civil engineering ; the aim was rather to find another exercise to which the Minimum Calculus could be applied. "M. Euler paraissait quelquefois ne s'occuper que du plaisir de calculer, et regarder le point de mécanique ou de physique qu'il examinait seulement comme une occasion d'exercer son génie et de se livrer à sa passion dominante."[1] Even Coulomb's pioneer work which will claim our attention later (page 146), carries a similarly revealing title, "Essai sur une application des règles de Maximis et Minimis à quelques problèmes de statique" (1773).

The principle of minimum effort leads, at least with Euler, to theological, metaphysical reflections. "As the design of the whole world is most excellent, and as it emanates from the wisest of all creators, there exists nothing in the world which does not reveal a maximum or minimum property. There can thus be no doubt that all effects in the world can be deduced just as well from their purpose, by determining the maxima and minima, as from the acting causes themselves."[2]

During the second half of the century, the theological aspect recedes more and more. Mach praises the penetration of rationalist thinking into the mechanical sciences with these words : "The humanistic, philosophic, historic and natural sciences come into contact with each other and are a mutual inspiration to free thought. Anybody who has, if only through perusing the relevant literature, re-lived this process of progress

---

[1] Condorcet in his "Eloge" of Euler, quoted according to E. Fueter, "Geschichte der exakten Wissenschaften in der schweizerischen Aufklärung". Aarau, 1941 ; page 260.
[2] Euler, 1744, according to Mach, page 436.

and liberation, will retain a lifelong sense of nostalgia for the eighteenth century."[1]

Today, we may be inclined to reserve our nostalgia for a different object. Mechanization, freed from any metaphysical connections, dominates human society, and achieves effects previously undreamt of, constructive as well as destructive ; but it has not brought about the Golden Age, envisaged by Rationalism. Instead of subscribing to Mach's praise of the liberation of mechanics from all inhibitions, we may be inclined to envy the religiously-minded scientists of the seventeenth and eighteenth centuries for the integrity of their conceptions ; for the confidence with which they regarded mathematics and philosophy, mechanics and theology almost as a single entity ; and for the piety which caused them to regard the domination of natural phenomena by unalterable laws as the manifestation of an intellectual principle.

[1] Mach, page 439.

CHAPTER IV

THE BUILDING TECHNIQUE OF THE
RENAISSANCE AND BAROQUE

1. THE ENGINEERS OF THE ITALIAN RENAISSANCE —
BUILDINGS, BRIDGES AND FORTIFICATIONS

In contrast to the Gothic style which owes its origin mainly
to structural problems, the Renaissance style must be regarded
as the offshoot of a new æsthetic ideal. The artists and builders
of the Italian quattrocento, however, showed a particularly
keen interest in technical problems. "The Experimenting
Masters" is the term which Olschki applies to the circle of
Brunelleschi, Ghiberti, Filarete, Leon Battista Alberti, Fra
Giocondo and others.

Filippo Brunelleschi (1377–1446), a pioneer of the new
architectural style, was not only a gifted artist but also an
accomplished technician who was actively interested in theory
and mathematics. His friendship with Toscanelli, the mathematician, can be regarded as one of the earliest examples of
collaboration between a technician and a scientist. The builders
of the fifteenth century were no longer content with their
traditional craftmanship. Arithmetic and geometry were placed
in the service of architecture and technique.[1] The "harmonious
proportion" of the buildings was to be determined by arithmetical and geometrical methods. In his surveys of ancient
Roman ruins, however, Brunelleschi was not only paying
attention to formal problems, "measuring the cornices and

[1] Olschki, I, page 36.

81

drawing the plans of those buildings".[1] He also studied the technical and structural details, "le cignature ed incatenature, e cosi il girargli delle volte . . . le collegazioni e di pietre e d'impernature e di morse".[1]

Vasari recounts how the master was, at an early stage, possessed of two fixed ideas : (1) to promote the advance of "good" architecture (i.e. that based on ancient examples), and (2) to vault the dome of Florence Cathedral (S. Maria del Fiore). His contemporaries were apparently even more impressed by his second achievement, which is essentially technical in character. This may, at least, be gathered from the fact that more than half of Vasari's entire biography is devoted to a description of the problems and difficulties in connection with the dome construction. It is somewhat difficult to reconstruct the details of this masterly engineering achievement from Vasari's rather anecdotic story, particularly in regard to the allegation that the vaulting was erected without centering. This method is praised as a special invention of Brunelleschi's which is said to have resulted in much greater economy than the proposals of the other "architettori ed ingegneri".

Apart from the æsthetic problems of reviving the architecture of Antiquity, the technique of vaulting remained one of the foremost preoccupations of the architects of the Italian Renaissance. In this matter, they went beyond the Gothic master builders and extended, for the first time, their interest to its theoretical aspects also. Leon Battista Alberti[2] was the first to propound a theory of the different kinds of vaults, though this theory was exclusively based on formal, geometric features. To him, the most perfect form of vault is the cupola which consists of a system of arches and, at the same time, of a system of rings ("cornici"), and thus possesses an excellent stability, permitting erection without centering.

If Alberti was the most important theoretician of the new technique, it is *Bramante* (1444–1514) who must be regarded as its greatest creative master. His design of the reconstruction of S. Peter's in Rome, as well as the execution of the great piers upholding the dome and their connecting arches of 141 ft.

---

[1] Vasari, "Vita di Filippo Brunelleschi".
[2] "De re ædificatoria", Vol. III, Chapter 14.

height and 79 ft. span, overshadow everything that had been achieved in the sphere of building construction since the days of antiquity. Vasari attributes to him a number of technical inventions, "invenzioni", e.g. a kind of "cast concrete" or artificial stone which was poured into previously prepared moulds ("invenzion nuova del fare le cose gettate . . . del buttar le volte di getto"). This technique, already known at the time of the Roman emperors, was used by Bramante, *inter alia*, for the erection of the great supporting arches of S. Peter's, mentioned above.

After the master's death, Giuliano da Sangallo, Fra Giocondo and Raphael were jointly entrusted with the supervision of the great building. Among these three, the one most akin to civil engineering was presumably the then octogenarian monk, *Fra Giocondo* (1433–1515), who may have been responsible for the strengthening and consolidation of the foundations which Bramante had carried out somewhat hastily.

The Dominican appears to have been particularly adroit in earthworks and hydraulic engineering. During the reconstruction of the Ponte della Pietra, Verona, he made the suggestion to protect the foundations of the mid-river pier, which stood on treacherous ground, by means of a deep-driven curtain of sheet piling.[1] He is said to have built, in the service of King Louis XII, two bridges over the Seine in Paris, including the Pont de Notre-Dame which, in accordance with the custom of the times, was flanked by "shopping parades" (like the Ponte Vecchio, Florence). According to Vasari, Fra Giocondo distinguished himself in the work of preserving the lagoon of Venice. The sedimentary deposits of the river Brenta were causing a progressive silting of the lagoon, the life artery of the town. To stave off the danger, a congress of Italy's most distinguished engineers and architects was convened. Vasari ascribes to Fra Giocondo the proposal to divert part of the Brenta water to the Lagoon of Chioggia, in order to restore the position at Venice and to put a halt to further silting. In fact, however, the Dominican's contribution was confined to four

[1] ". . . che detta pila si tenesse sempre fasciata intorno da doppie travi lunghe e fitte nell 'acqua" (Vasari, "Vite di Fra Giocondo e Liberale Veronesi").

reports which he submitted to the Council, whilst the actual work was entrusted to one Alessio degli Aleardi.[1]

Again according to Vasari, Fra Giocondo also submitted a design for the reconstruction in stone of the Rialto Bridge, at that time still a timber structure. Further designs for this most important bridge of Venice were later submitted by Sansovino, Vignola, Palladio and Scamozzi until, in 1587, the work was entrusted by the Senate of the Republic to *Antonio da Ponte* (*circa* 1512–1597), who is praised as having been "greatly experienced and versed in that art". Although the famous bridge, familiar to all visitors to Venice, has no unusual dimensions (93 ft. span ; 21 ft. rise), the technical details are of interest. The abutments, forming inclined layers of masonry, are adapted to the direction of the vault thrust, and the piling is staggered correspondingly. During the foundation work, the site was kept more or less free of water by the use of numerous pumps ("con l'uso di molte trombe").

When the foundations were already completed, their stability was called in question by sceptics. In particular, the responsible master was blamed for having used too short piles, or insufficiently driven piles.[2] An enquiry was held at which the master was able to prove that the piling had been driven in the appropriate manner. One witness testified to the fact that the piles had been driven in, until the penetration was no greater than 2 fingers after 24 blows.[3]

At that time, Italian engineers were famed for their technical skill far beyond the confines of their country. Quite a number were invited to take up appointments abroad. This happened, for instance, to *Aristotile Fioravante*, from Bologna (born between 1415 and 1420, died 1485 or 1486). In his capacity of

---

[1] Regarding the controversy between Fra Giocondo and Aleardi on problems of hydraulics, such as the relationship between flow, velocity and fall of rivers, see F. Arredi in *Annali dei Lavori Pubblici*, Rome, 1933, page 145.

[2] "Li palli erano curti, sottili et si battevano col battipalo da man, molto lesier . . . chi havesse voluto far uno buona palificata, bisognava mettervi migliori pali e addoperare un edificio da batter pali" (Arch. dei Frari, as quoted by Miozzi in *Annali dei Lavori Pubblici*, Rome, 1935, page 460).

[3] cf. E. Miozzi, "Dal ponte di Rialto al nuovo ponte degli Scalzi", in *Annali dei Lavori Pubblici*, Rome, 1935, page 450. This essay contains many interesting details and references to original sources concerning the construction of the Rialto Bridge.

municipal engineer of his home town, he accomplished, in 1455, the removal of a clock tower over a distance of approximately 50 ft., by means of special "machinery" invented by him. This engineering feat made his name famous throughout Europe. In the service of Duke Francesco Sforza, he built a bridge in the vicinity of Padua, together with Filarete. Later, he was in charge of canal and fortification works in Milan. In 1467, he worked for a time on the construction of a Danube Bridge in Hungary. He rejected a call to Constantinople, but accepted an appointment in Moscow in 1475.

His first work in the service of Grand Duke Ivan III was the reconstruction, in stone, of the great Cathedral of the Assumption in the Kremlin. It is said that his exact methods, using compass and straight-edge, commanded general admiration. Other churches followed later. In addition, he was active as a military engineer. In this connection, he was not only called upon to organize the casting of guns in a specially erected foundry, but also to erect a pontoon bridge during a campaign against Great Novgorod.

In 1485, he fell into disgrace with the Grand Duke because of his courageous defence of a wrongly accused German physician, and died soon afterwards.

Apart from bridge building, it was in the sphere of *fortress construction* that theory and practice, civil engineering, architecture and even the plastic arts were, during the Italian Renaissance, most intimately interwoven. "At a time when war itself became a matter of art and elegance, even the construction of fortifications had to be encompassed in the sphere of the beautiful."[1] This was in the spirit of Aristotle, who had stated that the walls of a town should not only be a protection but also an embellishment.

A typical example of the versatility and manifold interest of the masters of the Renaissance was *Francesco di Giorgio Martini* (1439-1502), a native of Siena and a military engineer in the service of Duke Federico of Urbino. A painter, sculptor and architect in his youth, he became municipal engineer of his home town and was, as such, responsible for the water supply.

[1] Burckhardt, "Geschichte der Renaissance", para. 108.

During his ten years at the Court of Urbino, he specialized in military engineering. Duke Federico was a well-known patron of the mathematical disciplines and their application to architecture and military engineering. To him, architecture was "based on arithmetics and geometry which belong to the most excellent of the seven free arts, because they show the highest degree of certainty".[1] His library included all the available works on arithmetics, architecture, ballistics and mechanics.[2] At this Court, Francesco became the "grandissimo ingegnere e massimamente di macchine di guerra" (Vasari) and received his main inspiration for his magnum opus, "Trattato dell' architettura". In this work, he deals with the entire subject of civil and military architecture, including the new defence and attack methods conditioned by the new firearms and explosives (angular alignment of mines galleries to avoid a rebound ; use of the magnetic compass for engineering purposes). Francesco di Giorgio is also regarded as the inventor of the polygonal or star-shaped bastions. The comparatively low "baluardi", with their base of enormous thickness and external batter, offer much greater resistance against artillery bombardment than the high, vertical walls of the medieval castles.

As a practical engineer, Francesco built a number of town castles, "rocche", for Duke Federico ; those of Mondavio and Cagli are still partially preserved. The mountain fortresses of San Leo and Sassocorvaro, north-west of Urbino, are also ascribed to him.

"Francesco's 'rocche' possess, quite aside from their military uses, a type of beauty which is akin to the functional beauty of certain modern industrial buildings. The fine relationship to site, as the architect makes the utmost use of whatever natural advantages the location may have, the frank expression of the innate qualities of the material he is using, the entire freedom from historical precedent, are everywhere observed."[3]

Francesco spent the last two decades of his life as a military engineer in the service of his home town, Siena. In this capacity, he was also called upon to carry out works of peace, such as

---

[1] cf. Burckhardt, "Geschichte der Renaissance", para. 31.
[2] cf. Olschki, I, page 127.
[3] A. S. Weller, "Francesco di Giorgio", Chicago, 1943, page 208.

Fig. 24. Bridge over the River Serchio, near Lucca   *Photo Alinari, Florence*

Fig. 25. Ponte del Castello Vecchio, Verona

Fig. 26. Star-shaped fortifications
From De Marchi, *Della Architettura Militare*

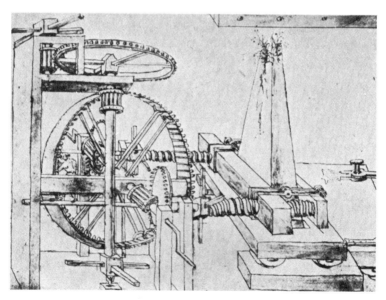

Fig. 27. Building machine　　　　　　　From *Skizzenbuch des
Giuliano da Sangallo*, published by Chr. Hülsen, Leipzig, 1910

the construction of dikes for a lake in the Maremma. On several occasions he was summoned to foreign towns and courts to advise on technical problems.

As an engineer, Francesco was an experienced practician. But, although he tried to apply mathematical methods to certain problems (e.g. ballistics), his mathematical training and scientific knowledge were far short of Leonardo's, who was about twenty years his junior. In other respects, the two men were somewhat akin, especially as regards their versatility and their fondness for a roving kind of life. (Regarding Leonardo, who was also active as a military engineer, see the preceding Chapter, page 55.)

Francesco di Giorgio Martini, Giuliano and Antonio da Sangallo, and Leonardo da Vinci were the last representatives of the epoch in which the construction of fortifications was entrusted to artists of general training.[1] The famous Italian military engineer of the sixteenth century, *Francesco de Marchi*, of Bologna (1504–1577), was no longer an artist but a professional soldier. His education was limited. In fact, according to one source, he is said to have been almost illiterate at the age of 32. Having been Artillery Inspector under Pope Paul III in Rome, he arrived at fortress engineering in the course of his military career. His engineering knowledge may have been furthered by his friendship with Antonio da Sangallo. During his Spanish service in the Netherlands, he earned such a high reputation that Philip II specially called for copies of his work "Della Architettura militare", for the training of the engineers and captains of the Spanish Army.

This famous and richly illustrated work was only published in 1597–1599, after the author's death. In the manner of the Renaissance, it is very comprehensive, with many historical excursions. No aspect of the subject is omitted : the origins and purpose of fortification works ; the knowledge required from a fortress engineer ; the details of design ; the execution and supervision of the work proper ; the building materials used,

---

[1] Among the German artists, it was Albrecht Dürer (1471–1528) who, like the Italians named above, concerned himself with geometry and the construction of fortifications. On the latter subject, he wrote a treatise : "Etliche Underricht zur Befestigung der Stett, Schloss und Flecken". Nuremberg, 1527.

etc. Even the contiguous spheres of hydraulic engineering and architecture are, to some extent, covered. Some of the general disquisitions of the author on the advantages and disadvantages of fortifications are still topical today, in the age of the atom bomb. On the strength of his ample experience, de Marchi concludes "that it is not advisable to fortify all cities, because too many fortresses cause great destruction in that they tend to lead to prolonged wars of siege. This is why wars last so long. In earlier times, when fewer places were fortified, the wars passed much more quickly, and were thus less costly to the sovereign in terms of money, and to the peoples in terms of destruction".[1] These very lines were encountered by the writer of the present book at a moment when they happened to be of tragic topicality : whilst the outcome of World War II was already beyond doubt, the German "hedgehog positions" continued to slow down the Allied advance and Hitler had just issued the order to fortify every town, every village, and every homestead in the mother country !

De Marchi was a forerunner of Vauban, whose innovations in the sphere of fortress construction partly go back to this Italian engineer, a century and a half earlier. The simple polygonal and star-shaped bastions introduced by Francesco di Giorgio now become those complicated designs of walls and moats, advancing and receding angles and empty spaces, "Cavallieri" and "Pontoni", which appear on so many old illustrations of towns (e.g. with Merian) and which, in the plan, often have a downright ornamental effect (cf. Fig. 26 ; also Fig. 33, Dunkirk).

Fortress engineering (in addition to hydraulic engineering) continued for a long time to be that sphere of engineering in which practice and theory, architecture and mathematics were most intimately interwoven. The military engineers, especially those concerned with ballistics, usually possessed a more profound knowledge of mathematics than, for instance, the

---

[1] ". . . non si conuiene far fortezze a tutte le Città . . . che (le fortezze) sono causa di maggior rouine, per li longhi assedij, che si tengono per acquistar detti luoghi forti. . . . Et questa è la causa che dura tanto la guerra in detti luoghi. . . . Ancora da pochi anni annanzi . . . et finiua più presto la guerra, et non era tanta spesa alli Prencipi, nè destruttione de Popoli" (First Book, Chapter XVIII, "Del fare la Fortezza alla Città").

architects of the Baroque who were mostly concerned with the building of churches. Moreover, many mathematicians and geometricians took an interest in fortress engineering, not only in Italy but also in France, Holland and Germany. This fact is borne out by such book titles as "Militärmathematik",[1] "Fortification und Messkunst",[2] "Problèmes de mathématique et de fortification"[3] and others.

2. THEORETICAL AIDS — MATHEMATICAL AND GEOMETRICAL RULES

A long time before the theories of statics and of the strength of materials were used to determine the dimensions of structures, certain empiric rules had been adopted for the design of frequently recurring building elements such as foundations, pillars, vaults, etc. These rules, condensed into concise mathematical or geometrical terms, were used by the master builders side by side with certain other rules of composition which were designed chiefly for æsthetic purposes.[4] At that time, we have to deal, not with the application of mechanics and statics to building construction, but rather with what one would nowadays call "rule of thumb". And yet, these rules may well be regarded as the beginnings of an "engineering science" proper in as much as they assumed a quasi-scientific character through the very fact that the results of accumulated individual experiences were formulated in mathematical or geometrical terms. Though not yet based on "statics", these rules do, after all, represent an application of scientific, if elementary, mathematics to practical building tasks. To some extent, these

[1] "Mathesis militaris, seu methodica calculandi, mensurandi, fortificandi et castrametandi ratio", by Gh. Meyer, Jena 1640.

[2] By Georg Schultze ; Erfurt 1639.

[3] By Pfeffinger ; 1684.

[4] Even the Greeks had attempted to formulate the principles of architectural composition in exact mathematical rules which were mainly based on certain recurring basic measures or proportions. (At the Temple of Ceres, Pæstum, the width measured at the architrave is thus exactly equal to the mean proportional of the height of the columns and the length of the temple ; cf. F. Krauss, "Pæstum", Berlin, 1941, page 40). The important part played by the square and the regular triangle in the design of the Gothic cathedrals, especially the façades, is well known. During the Renaissance, it was mainly Vignola who tried to base the entire "art of design" on the application of certain mathematical rules.

rules may be likened to modern "standard specifications" which have a similar purpose of standardizing solutions and dimensions which have proved their value in actual practice, and which exempt the craftsman and designer from the constant need to seek new solutions.

A number of such rules have come down to us from the architectural theoreticians of the Italian Renaissance. Leon Battista Alberti, for instance, stipulates the following standard dimensions for stone bridges in his "Ten Books on Architecture" ("De re ædificatoria". Florence, 1485) : The width of the piers should be one quarter of the height of the bridge ; the clear span of the arches should not be more than six times, and

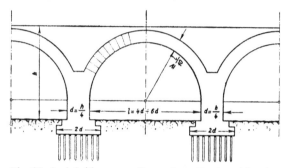

Fig. 28. Arch bridge, according to Leon Battista Alberti.

not less than four times, the width of the piers ; the thickness of the voussoirs should be not less than one-tenth of the span.[1] As Alberti was no doubt thinking of semi-circular arches, these rules result in a bridge elevation similar to that shown in Fig. 28. Some of the Roman bridges still intact are of a rather similar design, though most of them have comparatively stronger piers (frequently equal to one third of the span) ; they are thus less "bold".

For pile foundations, Alberti specifies a width of the piling, equal to twice the width of the wall to be carried. The length of the piles should not be less than one-eighth of the height of the wall to be carried, and their diameter not less than one-twelfth of their length.[2]

[1] Alberti, Vol. IV, Chapter 6.
[2] Alberti, Vol. III, Chapter 3.

Carlo Fontana (1634–1714), an architect hailing from the Italian-speaking part of Switzerland, prescribes the geometrical rules for the shape and dimensions of a masonry dome, illustrated by Fig 29. These rules, which are contained in his

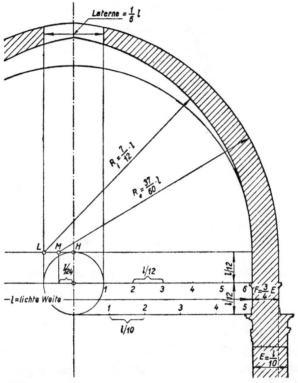

Fig. 29. How to determine the dimensions of a cupola, according to Carlo Fontana.

work, "Il Tempio Vaticano e sua origine" (Rome, 1694), are apparently inspired by Michelangelo's dome. According to these rules, the thickness "E" of the drum wall should be one-tenth of the clear span, and the thickness of the dome near the springings three-quarters of "E".

A further rule which has been widely applied, has been quoted by Bélidor in his "Science des ingénieurs", published in 1729 (cf. page 121), as well as by Viollet-le-Duc and even in our time, by Esselborn.[1] This is known as "Blondel's Rule" for the dimensioning of vault abutments ; it was applied to arches of any shape, and to abutment heights not exceeding 1.5 times the clear span. The rule is illustrated by Fig. 30 and

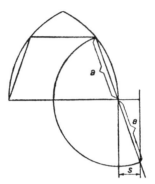

Fig. 30. Blondel's Rule to determine the dimensions of vault abutments.

may be roughly formulated as follows : An isosceles trapezoid is inscribed in the arch, and one side of it is extended downwards by its own length, "a". The vertical projection of "a" will then represent the thickness "s" of the pier or abutment.

3. MECHANICAL AIDS
BUILDING MACHINERY AND PLANT

The use of machinery and mechanical devices for the erection of major buildings has always been regarded as a feature of engineering construction. On the strength of this criterion, the Romans can indeed be regarded as *engineers*, as they made considerable use of mechanical devices for their major public works (cf. Chapter I, page 33).

There is, however, yet another reason why the building machines of the past must claim our attention here. As already pointed out (page 60), the observation of simple machines played an important part in the development of the science of mechanics. In modern times, theoretical knowledge is

[1] "Lehrbuch des Hochbaus", 2nd Edition, Leipzig, 1920 ; Vol. II, pages 242 and 243.

mostly ahead of its practical applications which are often inspired by scientific discoveries ; prominent examples can be found in the spheres of electro-technics (especially high-frequency) and nuclear physics. In those times, however, it was more common for scientists and mathematicians to derive, conversely, their inspiration from the practice inasmuch as they tried to fathom, scientifically, the operating principles of long-known mechanical devices. The Principles of the Lever and of the Inclined Plane, and in consequence, those of the Parallelogram of Forces and of the Statical Moments, formulated by the geometricians of the Italian Renaissance (cf. page 60), have been mainly inspired by the observation of the pulley, the windlass and the tackle, i.e., building plant and hoisting gear. Even Galilei, who belonged to a younger generation than Tartaglia, Cardano, Benedetti and Guidobaldo del Monte, gratefully acknowledges, in his "Discorsi e dimostrazioni matematiche", the inspirations which he owed to the observation of activities at the arsenal of Venice.[1]

For the design of hoisting devices, the technicians of the Renaissance made already occasional use of toothed gearing, sometimes in conjunction with worm gears, as is, for instance, apparent from drawings and sketches by Francesco di Giorgio Martini, Giuliano da Sangallo and Leonardo da Vinci. However, some of the examples contained in Martini's "Trattato dell' architettura" and thence copied by Giuliano da Sangallo[2] are so strange that they may have to be regarded as intellectual exercises rather than as actual designs. For these machines, composed of numerous toothed gears and worm gears with an aggregate gear ratio of several hundred (cf. Fig. 27), would no doubt have been much too complicated for the practical conditions on the building sites of those days. Most of the machines sketched by Leonardo have also remained on paper. But the individual elements depicted in these and other drawings

[1] "Largo campo di filosofare a gl'intelletti speculativi parmi che porga la frequente pratica del famoso arsenale di voi, Signori Veneziani, ed in particolare quella parte che mecanica si domanda ; atteso che quivi ogni sorte di strumento e di machina vien continuamente posta in opera da numero grande d'artefici . . . " (Galilei, First Day).

[2] " Skizzenbuch des Giuliano da Sangallo", published by Chr. Hülsen, Leipzig, 1910.

of contemporary artists, such as the screws and wooden spoke gears, etc., have no doubt been inspired by contemporary practice, though they may have been used on fixed installations (rotary cranes, mills, blowers in smelting houses,[1] etc.) rather than on building sites. Here, simple drums with a vertical shaft and four radial handles, turned by workmen or horses, seem to have been almost the only mechanical device in use, though they were sometimes combined with a kind of jib crane and tackle, as they are often shown on contemporary illustrations. Where necessary, the number of such drums could be increased *ad libitum*, as was done by Domenico Fontana during the erection of the obelisk on S. Peter's Piazza, Rome (cf. Fig. 32, page 100).

The power for the working of cranes and hoisting gear was supplied, throughout the Middle Ages, by the same treadwheel which was already known to the Romans (cf. page 34 and Fig. 13). The circular windows representing a so-called "wheel-of-fortune", i.e. a spoked wheel encircled by human figures, occasionally encountered in Romanesque architecture (e.g. in Beauvais Cathedral, and above the St. Gallus Portal at the North Transept of Basle Cathedral) are thought, by some, to be emblematic versions of the treadwheel.

Apart from transport and hoisting devices, it is the *hydraulic* machines which have continued in use throughout the ages. In ancient Egypt, swinging-booms and bucket wheels were used for irrigation purposes, probably supplemented during the last centuries B.C. by water screws ("Archimedes' Screws", cf. page 24) and, occasionally, by piston pumps. The Romans made use of bucket elevators and water screws for the drainage of foundation pits in bridge building and hydraulic engineering, as well as for pit drainage. During the Middle Ages, the Renaissance and frequently right up to modern times, the devices used for the same purpose were rather similar. A hydraulic machine which is rather complicated for the early fifteenth century, is shown on a drawing of the Veronese painter, *Pisanello* (approximately 1395–1455). It consists of a bucket elevator which is driven by an undershot waterwheel by means

---

[1] For illustrated examples, see Beck, Vol. II, page 131.

of toothed gearing. A second machine, depicted on the same drawing, represents a kind of piston pump, built of wood and driven by the river itself by means of a water wheel and connecting rod.[1]

A brief reference to *topographical instruments* is relevant in this connection. It is perhaps chiefly in the sphere of surveying and marking out for subsoil and hydraulic engineering works that the close liaison of science and practice, which is now regarded as typical for engineering works, was realized first.[2] The history of mathematics begins with the methods used thousands of years ago by the "agrimensors" of Egypt and Mesopotamia to solve their agricultural and irrigation problems (cf. Chapter I, page 2).

Vitruvius describes the instruments used by the builders of the Roman aqueducts to achieve the requisite uniform minimum fall of the flights of walled arches which carry the water across the Campagna : the "dioptra" and the "libra aquaria" (water level). Up to the Italian Renaissance, e.g. in the days of Leon Battista Alberti, the number of instruments in use had hardly increased. In the course of the following centuries however, the surveying methods were subject to a continuous process of development and refinement. In this process, Swiss surveyors and engineers played a prominent part. The problem of measuring distances and heights by indirect methods was solved by means of the elementary theorems of the triangle. "A new geometric instrument or triangle permitting to measure all dimensions of height, width, length and depth easily and without calculation", was manufactured and sold by *Leonhard Zubler* (1563–1609), a Zürich goldsmith and mechanic. Together with a Zürich mason, *Philipp Eberhard*, Zubler was also the first to carry out field surveys with plan tables. This was even before (about 1600) Professor *Johannes Prætorius* of Altorf University pointed to this simple graphic survey

[1] Reproduced as Fig. 147 of the Italian edition (Turin, 1945) of Degenhart's "Pisanello".

[2] Jakob Bernoulli wrote in 1684 : "The art of surveying can rightly be exercised only by somebody who is experienced in mathematics. Contrary to a strange prejudice, it should therefore not be entrusted to uneducated and ordinary citizens" (according to E. Fueter, "Geschichte der exakten Wissenschaften in der schweizerischen Aufklärung". Aarau, 1941 ; page 61.)

instrument (since named "mensula prætoriana" after him) which permitted the immediate *in situ* recording of field surveys in the form of plans (cf. Fig. 31).[1]

4. A "CIVIL ENGINEER" OF THE SIXTEENTH CENTURY — DOMENICO FONTANA[2]

An example may convey an impression of the activities of a civil engineer of the sixteenth century. If we select *Domenico Fontana* (1543–1607) for this purpose, this is only because the author happens to be particularly well acquainted with his activities which are, however, typical for many another master of his age.

Let us begin by pointing out that, even during the sixteenth century, there is as yet no sign of any professional differentiation between "architect" and "civil engineer" proper. It is only on the strength of his personal ability and inclination that one master may be regarded chiefly as an artist and architect, and another more as a technician and engineer. Fontana, too, was an architect and, in this capacity, erected a number of important palaces in Rome and later, at Naples. In contrast to other contemporary masters, however, Fontana has a special claim to be regarded as a *civil engineer*, for the following reasons :

(1) His ingenious and confident mastering of *difficult technical problems*, such as the erection of the great obelisks in Rome ;

(2) His extensive occupation with *road and hydraulic engineering works ;*

(3) His interest in the details of *practical building construction* or "site organization" as we would call it today ;

(4) His knowledge of mathematics and geometry, and his deliberate use of *calculations* to check intuitively chosen solutions of mechanical problems.

---

[1] cf. a contribution by Leo Weisz to *Neue Zürcher Zeitung* of 12th and 19th December, 1943, Nos. 1995 and 2060.

[2] This section is partly reproduced from an article by the author which has appeared in *Schweizerische Bauzeitung*, Vol. 123, page 172.

The removal of the great Egyptian obelisk which, since the days of ancient Rome, had stood by the side of S. Peter's on the site of the Circus Maximus, and its re-erection in the centre

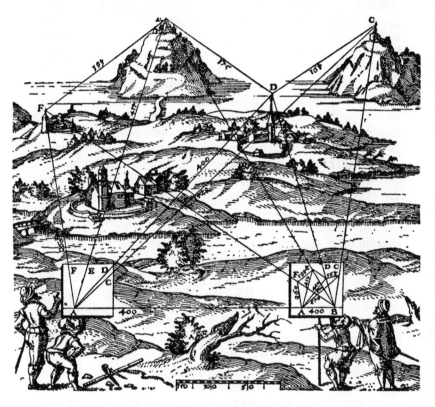

Fig. 31. Field survey according to Zubler, 1607.
From L. Weisz, *Die Schweiz auf alten Karten, Zürich.*

of the great piazza in front of the new church, represented an amazing technical achievement which commanded the admiration of the contemporary world. The successful completion of this feat was commemorated by specially coined medals. Fontana has described it in great detail in a *de luxe* publication, embellished with many engravings and entitled : "Della trasportatione dell'obelisco Vaticano et delle fabriche di nostro Signore Papa Sisto V, fatte dal Cavallier Domenico Fontana, architetto di Sua Santità."

A competition had been organized to obtain the best solution for the job. The committee was confronted with a great variety of proposals envisaging the lifting of the enormous monolith by means of tackles, pulleys or levers, and its transport in a vertical, horizontal or even oblique position. Fontana's proposal, which was illustrated with a small-scale model of lead, was deemed to be the best, and was accepted.

The original intention was to entrust the supervision to two experienced older architects, Bartolomeo Ammannati and Giacomo della Porta. But Fontana pointed out that the responsibility for such a difficult and risky undertaking should not be divided, as the supervisory engineers might blame a possible failure on to the author of the project, and vice versa. Eventually, Fontana was put in sole charge, at the age of 42 years.

The preparatory work took seven months. Tremendous quantities of timber, iron, ropes, windlasses, tackles, etc., had to be assembled, and a particularly robust scaffolding (Fig. 32) had to be erected. On 30th April, 1586, when everything had been prepared down to the smallest detail, a beginning could be made with the first phase of the work, lifting the obelisk from its base and placing it in a horizontal position. The operation and the communication of orders were planned with extreme minuteness. A blow of the trumpet called for the operation of the windlasses, a bell signal for their stoppage. Each windlass and the corresponding ropes were numbered so that each of them could be regulated individually or in groups. To avoid disturbances, the most stringent disciplinary measures —including the death penalty—were taken. A public executioner was, in fact, ready on the site.

For the transport of the great monolith and its re-erection on a new, lower level, an earth wall was built and reinforced with timber so that the load could be moved on rollers almost horizontally. The new base, prepared in advance, had to be founded on piling on account of the poor subsoil. Around the base, the earth dam was enlarged so as to form a kind of working platform. Here, it was no longer necessary to lift the obelisk vertically, but merely to turn it by 90° and to place it on the four supports, representing bronze lions (cf. Fig. 32).

Among other achievements establishing Fontana's fame as an eminent technician and engineer, may be mentioned the erection of three further obelisks, including Rome's largest, that at the Lateran Basilica ; the transport of a small oratory dating from the thirteenth century which, encased in a solid timber cage, was transported *as a whole*, without being dismantled, to the newly built Sistine Chapel of S. Maria Maggiore; the erection of the statues of the Apostles Peter and Paul on the columns of Trajan and Marcus Aurelius, respectively, and the restoration of the last-named column. As a collaborator of the Cathedral architect, Giacomo della Porta, Fontana was also concerned with the most prominent building task of his time, the vaulting of S. Peter's cupola. *Michelangelo* (1475–1564) had completed the drum of the great dome up to the height of the upper cornice, and had left a model of the dome. After the master's death, the construction of the dome was deferred, and it was only some twenty years later that *Giacomo della Porta* (approximately 1537–1602 ; since 1573 Architect of S. Peter's) and Domenico Fontana undertook the daring task of vaulting the dome, giving it a slightly higher shape than Michelangelo had intended. To absorb the horizontal thrust, three iron rings were built in. These did not, however, succeed in preventing some damage and cracks which gave rise to thorough examinations and restoration works a century and a half later (cf. Chapter V, Section 2).

Fontana also began the construction of a bridge carrying the Via Flaminia across the Tiber near Città Castellana, but he was not able to complete this work, as Pope Sixtus V died during the construction, and Domenico had to give up his office as papal engineer and architect.

In addition, Fontana was entrusted with town planning tasks of some magnitude. Even today, the structure of the Eternal City still bears the imprint of the great streets planned by him, connecting distant quarters in a straight line, such as the Via Sistina and its extension, the Via delle quattro Fontane, the Via Panisperna and others. His aqueduct, called Acqua Felice after the Pope's temporal name, permitted the utilization of

Fig. 32. Erection of the obelisk on St. Peter's Piazza, Rome.
From Domenico
Fontana, *Della trasportatione dell'obelisco Vaticano . . . Rome*, 1590.

the high-lying quarters, such as the Quirinal and Esquilin, for residential purposes.

As a road builder and hydraulic engineer, Fontana anticipated one feature of civil engineering which is about to assume greater importance in our times, namely the landscaping activities of the engineer. As will be shown in the following chapters, the development during the eighteenth and nineteenth centuries tended to render science and research subservient to the engineer, and to assign to him, more and more, the task of an abstract calculator. It is only in recent years that greater emphasis is again laid on the fact that great engineering works, such as roads, bridges, dams, canals or reservoirs, are of profound influence on the external appearance of the circumambient landscape, and that such objects must, from the outset, be conceived as an *entity*, having full regard to all the aspects of the task — practical, economic and æsthetic.[1] It is in this spirit that Fontana has approached the town planning tasks with which he was confronted. His great new streets display a supreme unity of utilitarian and æsthetic considerations. They fulfil the Pope's wish to create short and convenient links between the different basilicas and districts, and they comply, at the same time, with the æsthetic requirements, so dear to the Baroque, of creating distant view-points and large, integral spaces. So well achieved is this duality of purpose that there is no point in putting the question whether we have to deal with merely utilitarian works, or with works of architecture.[2]

It was the same endeavour to combine functional purposefulness with æsthetic appearance that led Fontana, during the construction of the great aqueduct, to the happy idea of "mostra", i.e. of *demonstrating* the water in a monumental fountain, after it had been ducted, with great effort and technical ingenuity, over a distance of 15 miles, and before it was turned to its practical purpose of water supply and irrigation.

---

[1] In this connection, see Alwin Seifert's book, "Im Zeitalter des Lebendigen"; Munich, 1941.

[2] Fontana's *engineering* ability is again manifest from his care for a solid street pavement which he tried to find through a series of tests with different kinds of pavement. It is said that during the first half of the year 1587 alone, no less than 121 streets of Rome were paved, under his direction as papal architect.

The idea was widely applauded and has subsequently been copied in the case of other Roman aqueducts, e.g. that of the Acqua Paola, built by Domenico's brother, *Giovanni Fontana*.

During his subsequent activities at Naples, Domenico Fontana was likewise concerned with extensive hydraulic works and road construction. He was also entrusted, by the Viceroy, with the construction of a new harbour basin. But his grandiose plan was only carried out after his death, and then only partially.

In the case of many of these works, not inconsiderable difficulties had to be overcome, and it is typical for Fontana's *engineering* qualifications that he concerned himself with all the practical details of the work, sometimes from a contractor's point of view. Among the problems in connection with the transport of the Vatican Obelisk, those of procuring the requisite materials figure prominently in his report. The work was carried out "by direct labour", to use the modern term. The Pope placed generous funds at the disposal of his engineer, and conferred on him far-reaching powers of expropriation, etc.

In describing the great works carried out by him, Fontana makes a habit of stating the number of workers engaged, and often also their allocation to different individual operations. Thus, 300 men were required for the pit drainage alone when the Lateran Obelisk was removed from its subsoil pit in the vicinity of the Tiber. The description also covers other technical and administrative details, such as the safety measures adopted for the protection of the workers. For instance, during the lifting of the Vatican Obelisk, the carpenters endangered by falling objects had to wear iron helmets for protection.

In connection with the subject matter of this book, it is mainly of interest to determine whether Fontana, in all his difficult tasks, was simply relying on working experience, routine and intuition, or whether, and to what degree, his schemes were of an *engineering* character in the modern sense, in that he deliberately relied on calculations. Though self-taught, he had acquired, in his youth, a good knowledge of geometry, as is pointed out by his biographers of the seventeenth century (Baldinucci, Bellori). He was a contemporary of the Italian Renaissance mathematicians and geometricians, Tartag-

Fig. 33. Port and fortress of Dunkirk   From Bélidor, *Architecture hydraulique*

Fig. 34. Vignette with the attributes of the bridge builder : pile driver, levelling instrument, drawing, winch, crown with the Fleur-de-Lis. In the background, fragments of a building and a stone bridge (pier of the Pont de la Concorde)

From Perronet, *Déscription des projets . . .*

Fig. 35. Bridge building site with de-watering machinery

From Perronet, *Déscription des projets . . .*

lia, Cardano, Benedetti, Guidobaldo who (as mentioned on page 59) concerned themselves with problems of mechanics, and formulated the first mathematically expressed theorems of statics. It is not known, and, in fact, rather doubtful whether the busy papal engineer managed to find the time to study the new books on mechanics, just published. But there is no doubt that he mastered the basic principles of contemporary science.

Perusing Fontana's "Della trasportatione . . .", we must, of course, not look for a proper structural analysis. But our expectation will not be wholly disappointed. As an engineer *par excellence*, he did indeed carry out preliminary theoretical calculations, in keeping with the advance of the exact sciences of his time. To determine the number of ropes and windlasses required to lift the monolith, it was first necessary to determine its weight. If we follow Fontana's calculation which, to him, appeared important and novel enough to be assigned two large pages of his work, it is interesting to note the absence of so many elementary aids which would, today, enable any school-boy to solve the problem in five minutes. To determine the volume of the truncated pyramid, he does not use the modern formula but elaborately dissects the body into its elements : the prismatic core, the four wedge-shaped lateral pieces, and the four corner pyramids, the volumes of which are calculated separately and then totalled. Instead of using decimal fractions (which were first introduced by Rudolff during the sixteenth century and gained notoriety mainly through Simon Stevin's "Arithmétique", which appeared in 1585), Fontana was compelled to work laboriously with vulgar fractions. By weighing a cube of one span side length, the specific gravity of the stone was found to be 86 Roman Pounds per cubic span (approximately 160 lb. per cu. ft.). The weight of the entire obelisk was thus calculated at 963,537 35/48th Roman Pounds (approximately 327 long tons). On the strength of this calculation, Fontana fixed the number of tackles and windlasses to be used at 40. In addition, he provided for the use of five huge levers.

No other theoretical calculations are contained in Fontana's report, and it is hardly likely that he carried out calculations of, say, the tensile forces of the pulley, windlasses and tackles as these were well known from experience.

Fontana may be regarded as typical of the many master builders who, even during the sixteenth and seventeenth centuries, combined the functions of architect and engineer, artist and technician. The scope of mechanical knowledge required by the practician was still confined to the sphere of elementary, immediately evident relationships and laws. As before, the study of mathematics and geometry on the part of architects and engineers served, not so much to solve structural problems directly by means of calculation, but rather as a means of education, in order to develop the power of three-dimensional and mechanical imagination. As in the days of the Gothic age and of the early Renaissance, the structural engineer was still more of an artisan and artist than a scientifically-minded specialist.

CHAPTER V

THE ADVENT OF "CIVIL ENGINEERING"

1. KNOWLEDGE OF BUILDING MATERIALS— FIRST TESTS OF THEIR STRENGTH

The knowledge of the properties of building materials has gradually developed into an important branch of engineering science. In this development, three phases can be distinguished which may be called the artisan stage, the descriptive stage and the quantitative stage. (As we are solely concerned with *engineering*, no mention shall here be made of those modern scientific and analytical testing methods which are more the concern of the chemist and physicist than that of the engineer).

The carpenter or the mason may well have acquired, through years of experience, a thorough knowledge of the properties of the material which he handles. This is borne out by the buildings of the past, especially by modest utilitarian buildings which are often made particularly attractive through the expert handling and treatment of the materials used. But even where the strength of the material is exploited to the limit, as is the case with the Gothic vaults and buttresses, the artisan's knowledge of that strength is largely instinctive in character. It is, in fact, this instinctive approach which marks the difference between artisan and engineer.

A *scientific* approach to the problems of building materials is first apparent in the works of the architectural theoreticians who, following the example of Vitruvius (mainly Book II), deal at great length with the properties of wood, natural and artificial stone, binding agents, etc. *Leon Battista Alberti* (Vol. II, Chapters 4–7) classifies the different species of wood in accord-

105

ance with their suitability for different purposes : alderwood is unsurpassed for pile foundations in water-logged soil but not sufficiently durable when exposed to the air and the sun. Elmwood hardens in air but is otherwise liable to crack. The wood of the spruce and stone pine is suitable for underground works but is liable to shrink and twist when exposed to the air, and to be attacked by sea water, which is not the case with olive and oakwood. The latter has unlimited durability. But the best building material is supplied by the firtree ; its trunk is straight, easily workable and comparatively light, the only drawback being its liability to catch fire easily.

True to the manner of the Renaissance, Alberti draws his knowledge, to some extent, from antiquity, or quotes ancient authors as witnesses for his own observations, e.g. Theophrastus (a pupil of Aristotle, *circa* 372–287 B.C.), Vitruvius and others, expressing his astonishment at the fact that they do not have a higher opinion of the walnut tree. He discusses at great length the most favourable cutting season and, somewhat more cursorily, the different properties of the heartwood, sapwood, roots, etc.

In his analysis of stone material (Vol. II, Chapters 8 and 9), Alberti is less methodical. Italy's most common kinds of stone are merely listed geographically. One chapter contains general advice to the builder : stone with numerous veins is liable to fissurize and will not last long ; reddish and ochraceous veins being the most dangerous ones. Newly quarried stone is softer than seasoned stone ; moist material can be more easily cut than dry material. If submersion in water results in a considerable increase in weight, the stone will decay through the influence of humidity. Stone that does not resist the flame is dangerous in case of fire.

In this connection, Alberti quotes Cato and Tacitus but points out, rightly, that the value and durability of the different kinds of stone can better be assessed from the observation of ancient buildings than from the writings of the philosophers.

For the manufacture of bricks, Alberti recommends the use of whitish or reddish clay which contains not too much sand. To avoid frost danger during the winter, the bricks should be stored in dry sand. To avoid the risk of premature drying during the summer, they should be covered with moist straw. To achieve uniform burning, the bricks should not be too thick, or else have holes extending to half the thickness.

Dealing with binding agents, Alberti describes the raw materials, the process of manufacture and the subsequent treatment (slaking, storage) of lime and gypsum. Among the aggregates needed for mortar making he mentions pit sand, river sand, and sea sand, preferring sand from pits or from declivous mountain streams. The sand must, in any case, be granular and free from earthy admixtures.

In the same purely descriptive manner, other theoreticians of the Renaissance have also dealt with building materials in their works on architecture. In his work on fortifications, F. De Marchi (cf. page 87) deals at length with building materials like timber, stone and binding agents, quoting extensively from Greek and Latin sources and also from Alberti, whose very phraseology is often apparent.

Vasari (1512–1574) discusses the most usual kinds of stone in the introduction to his treatise, "Lives of Artists".[1] As he ignores the literary sources of antiquity, his presentation of the subject matter has a more modern, scientific appearance. Whereas Alberti and De Marchi classify the different types of stone merely according to geographical principles, or according to their colour shade, Vasari applies petrographical criteria and presents separate descriptions of porphyry, serpentine, granite, the different kinds of marble, travertine, slate, etc.

But it was only during the third stage of development that the science of the building materials became the indispensable and efficient aid of the engineer. That was when their *strength properties*, which are of vital interest to the engineer, were made the subject of research and systematic observation. After Galilei had introduced gauging and testing into physics, and Francis Bacon, Lord Verulam, had called for the inductive method of scientific inquiry, it did not take long, until the elasticity and ultimate strength of solids were also made the subject of quantitative analysis. Again, the initiative came from physicists who were the first to carry out such tests ;

[1] Chapter I, "Delle diverse pietre che servono agli Architetti".

but they did so without thinking of their practical application, least of all in the sphere of building construction. This fact is, for instance, borne out by the comparatively important part played by *glass* among the materials tested by the scientists of the seventeenth and early eighteenth centuries. This material is particularly well suited for the observation of elasticity and strength. But it was, at that time, still far from being an important building material.

Among the earliest strength tests[1] are those of *Mersenne* (about 1626) as well as those carried out by *Mariotte*[2] (about 1670–1680) with wood, metal and glass rods subjected to tensile and bending stresses. The results confirmed Galilei's theory according to which the bending strength must be proportional to the square of the cross-section height. But they were — understandably — at variance with Galilei's thesis regarding the ratio of bending and tensile strength (cf. page 65). At the same time, *Hooke* investigated the behaviour of elastic bodies, especially spiral springs, and thereby discovered the law named after him (cf. page 69). As already mentioned, Hooke's tests also extended, from the outset, to metals, stone, glass, etc.

During the eighteenth century, known as the age of encyclopedia and general comparative description of nature, the elasticity and strength of a great many materials were tested and compared in extensive series of tests. In 1707 and 1708, *Parent* published the first tabular juxtapositions concerning bending tests with beams of oakwood and deal. We must also mention the tests made by *Girard* ("Experiences pour connôitre la Resistance des Bois de Chêne et de Sapin", Mém. Acc., Paris, 1707) and by *Réaumur* ("Experiences pour connaître si la force des cordes surpasse la somme des force des fils que composent ces mêmes cordes", Paris, 1711).

Complete and accurate tables showing the ultimate strength of different kinds of woods, metal, glass, etc. subjected to compression, tension and bending, were published in 1729 by the

---

[1] True, a sketch by Leonardo da Vinci (Cod. Atl. Sheets 152 and 211) shows a beam on two supports, carrying a concentrated load (weight) in the centre. But it is hardly possible to decide whether this sketch refers to a test actually carried out, or simply reflects a mental exercise.

[2] Regarding these tests and others mentioned later, see (*inter alia*) Saint Venant, para. 61.

physicist *Musschenbroek* (1692–1761) of Leyden in a treatise, still written in Latin and entitled "Introductio ad cohaerentiam corporum firmorum". In this connection, Musschenbroek observed for the first time (15 years before Euler's theoretical derivation of the buckling formula) that the buckling resistance of slender struts decreases with the inverse ratio of the square of their length.[1] It was on Musschenbroek's figure for the strength of iron that the first scientific attempt was based, in 1742, of calculating the dimensions of a building element as a function of the statical load (cf. the following section, page 111 *et seq.*).

Numerous strength tests, mainly bending tests, were also carried out by that famous naturalist, *Comte de Buffon* (1707–1788), author of the well-known "Histoire Naturelle" and keeper of the Jardin des Plantes, Paris. He tested iron rods and, mainly, wooden beams, varying their quality, the cut and age of the wood, and the cross-section and length of the beam within wide limits (cross-section varying between 11×11 centimetres and 22×22 centimetres ; length varying between 2.30 and 9.10 metres). Buffon was the first to measure the sag prior to rupture. In the case of simple beams carrying a concentrated load, he found, like Galilei, that the ultimate strength was approximately proportional to the width of the cross-section and to the square of its height, and inversely proportional to the span.[2]

A knowledge of the strength coefficients of the more important building materials was the indispensable pre-requisite for any practical application of the theorems of statics to practical structural tasks. The test results and strength coefficients published by scientists like Parent, Musschenbroek, Buffon and others were destined to play a great part later on. As will be shown in the following sections, it was during the second half of the eighteenth century that the science of engineering proper came into existence, and with it the modern civil engineer who based his designs on scientific calculation.

---

[1] Musschenbroek thereby assumes $P = k\frac{d^3}{l^2}$, whereas the correct formula for slender struts (within the range of the Euler curve) is $P = k\frac{d^4}{l^2}$.

[2] cf. Saint-Venant, paragraph 61.

The technicians of that time not only derived considerable benefit from the elder scientists' pioneer work, they also carried out extensive tests of their own, closely related to the great engineering works of the times.

It was still during the eighteenth century that the French engineers, *Bélidor* and *Perronet* (cf. page 121 *et seq.*) published their tables of strength coefficients, and *Aubry*, Inspector-General of the French Bridges and Highways, carried out and published (1790) his tests on oakwood beams. Among the most

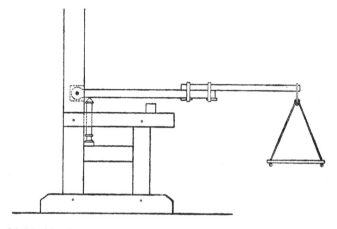

Fig. 36. Machine for stone cube crush tests.
From Gauthey, *Traité de la construction des Ponts.*

important pioneering efforts, however, were the extensive tests carried out by *Gauthey, Soufflot* and *Rondelet* at the time of the construction of S. Geneviève's Church (Panthéon), Paris, and later when, after the removal of the scaffolding, cracks became apparent in that building and an investigation was made. They included the first compression tests ever made on stone and mortar, and interesting bending tests with iron-reinforced stone beams.[1] Fig. 36 shows the test apparatus used by Gauthey. The compression force required to crush the stone cubes was produced by means of weights, acting on a lever arrangement.

[1] cf. Gauthey, Rondelet, as well as *Génie Civil* 1930, page 189.

2. STATICS APPLIED TO PRACTICAL CONSTRUCTION — STATICAL ANALYSIS OF ST. PETER'S DOME, ROME — POLEMICS BETWEEN THEORETICIANS AND PRACTICIANS[1]

It was about the middle of the eighteenth century that the first attempt was made to apply the methods of exact science to a practical building task. The principles of statics had already been formulated. In particular, the energy equation, i.e. the Principle of Virtual Displacements, was already known to numerous mathematicians and technicians. The strength properties of some important building materials, such as wood and iron, had been investigated, and tables showing the results of bending and breaking tests had been published. It was in keeping with the mentality of the rationalist age that the statical behaviour of structures, hitherto designed merely according to architectural feeling, should also be subjected to the searching control of reason, and that their dimensions should be determined in accordance with the rules of mechanics which was now emerging as a science. The change from artisan routine to modern, scientifically based civil engineering must be regarded as truly revolutionary. It marks the beginning of a unique development which has not yet reached finality. It is therefore not out of place to discuss, in some detail, one of the earliest examples of this new mentality. During the years 1742 and 1743, at the behest of Pope Benedict XIV, a structural analysis was made of S. Peter's dome, in order to ascertain the cause of the cracks and damage which had become apparent, and to devise remedial measures. In 1743, a printed report on the subject was published under the title "Parere di tre matte-matici sopra i danni che si sono trovati nella cupola di S. Pietro sul fine dell' Anno 1742".

The three authors, Le Seur, Jacquier and Boscowich,[2] were

---

[1] Part of this section is reproduced from an article contributed by the author to *Schweizerische Bauzeitung*, Vol. 120, page 73 (15th August, 1942).

[2] Regarding Boscowich, cf. page 159. Jacquier (1711-1788), in his time an outstanding figure of the "Roman Republic of Scientists", was also in contact with Winckelmann, Angelica Kaufmann and Goethe. The latter writes, on 25th January, 1787 : "Vor einigen Tagen besuchte ich den Pater Jacquier, einen Franziskaner, auf Trinità de' Monti. Er ist Franzos von Geburt, durch mathematische Schriften bekannt, hoch in Jahren, sehr angenehm und ver-ständig. . . ." (Goethe, Italienische Reise).

conscious of the fact that they were breaking new ground. After an apology, addressed to practicians and building experts,[1] they point out that the building concerned is unique in the world. There may be sufficient experience available to deal with minor structures. In this exceptional case, however, the position could not be properly assessed without going back to theoretical, mathematical reflections.

There follows a regular, detailed survey first of the dimensions and design of the building, and subsequently, of the damage observed at different times. The report goes on to mention, and to reject, as unfounded, various possible explanations such as a subsidence of the foundations, or a weakening of the piers through the recesses or spiral staircases. Finally, a yielding of the impost ring of the dome is named as the true cause of the trouble. And now begins the second, more interesting part of the report. An attempt is made to calculate the horizontal thrust, and to prove that the two[2] tie rings, built into the dome at the time of its erection, were no longer able to withstand that thrust. The method followed by the authors is of considerable interest. Instead of using a force polygon, as modern thinking might have led one to assume, they used the method which was first mentioned by Jordanus and his followers and later refined and generalized by Descartes and Johann Bernoulli,[3] and which is now known as the "Principle of Virtual Displacements".

The authors then present a graph showing how, according to their findings, the dome must have given way. They regard the observed fractures as a kind of moving joint or hinge around which the intact parts of masonry have turned (cf. Fig. 37). These parts are regarded as rigid between the fractures. A kind of energy equation is used to equate the work of the sagging or rising heavy masses (the latter case occurs, for instance, when the abutments commence their tilting movement) with the energy, still inadequately understood, of the

---

[1] "Saremmo forse anche in obbligo di scolparci presso que' molti, che non solo preferendo la pratica alle teorie, ma stimando quella sola necessaria ed opportuna, e queste forse ancora dannose" ("Tre mattematici", page 4).

[2] According to more recent research, there were three such rings (cf. page 99 ; also Beltrami, "La Cupola Vaticana").

[3] cf. pages 68 and 69.

deformation of the horizontally expanding iron tie rings. The ratio of the relative displacements is obtained geometrically from the graph.

But, even apart from the unrealistic assumption that the intact parts of the masonry, between the visible fractures, can be regarded as geometrically rigid, the calculation is not quite correct. True, the individual weights are accurately determined on the basis of the specific gravity of the different building materials. But the authors confuse the notions of virtual and

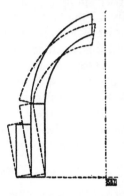

Fig. 37. Graphic illustration of displacements. From *Parere di tre mattematici.*
Reproduced from *Schweizerische Bauzeitung*, Vol. 120.

elastic displacements, as they are still wholly in the dark as to the character of elasticity. They do, it is true, recall the fact that iron will expand, and they quote the French physicist Philippe De la Hire and others as having observed the expansion of the metal through the effect of solar heat. The fractures observed on the building are thus connected with the elongation of the iron, caused by the thrust.[1] But the resistance of the tie rings (which grows with the elastic elongation) is thereby erroneously treated as a *constant* force and simply added to the tilting resistance of the piers. An elastic value is thus bracketed with a static value, which is inadmissible.

---

[1] "Quell 'allungamento, che in poco tempo cagiona il caldo o del Sole, o del fuoco, lo deve qui aver prodotto in più d'un secolo e mezzo, l'azione continua di una spinta così gagliarda" ("Tre mattematici", page 20).

This error is typical for the notional difficulties which had to be overcome during the genesis of statical thinking. It emphasizes, once more, the eminent importance of Hooke's achievement for the development of engineering (cf. page 69).

In detail, the course of the calculation is approximately as follows : The weight of lantern and cupola exerts, on the impost ring, a total thrust (spread over the entire circumference) of $H = \Sigma \, G \, \frac{v}{h}$, where G denotes the weights, v the sag of the centre of gravity, and h the corresponding horizontal displacement of the impost, individually for the lantern and the different sectors of the cupola. The thrust H is opposed by the resistance, W, which consists of the tilting resistance, similarly determined, of the drum wall and buttresses, and the resistance of the iron tie rings. The latter is calculated from the cross-section of the rings and from the ultimate strength of iron, taken from Musschenbroek's work (cf. page 109) — 600 lb. for a wire of 1/10 in. diameter. In this connection, the position of the individual rings and the relation of radial thrust, p, and longitudinal tensile force, Z, are taken into consideration : $Z = pr = \dfrac{H.}{2\pi}$

The deficiency in horizontal resistance at impost height is calculated at 3,237,356 Roman Pounds (approximately 1,100 tons) which it is recommended to remedy by the use of additional tie rings. A safety factor of 2 is taken into consideration ; this fact, too, may be regarded as a symptom of true engineering mentality.[1]

The method used by the three mathematicians, and the results at which they had arrived, immediately aroused adverse criticism from many sides. The novelty of the procedure, the application of mathematics and of mechanical theorems to the survey of the stability of the dome met with the opposition of the practicians. "If it was possible to design and build S. Peter's dome without mathematics, and especially without the newfangled mechanics of our time, it will also be possible to restore it without the aid of mathematicians and mathematics . . . Michelangelo knew no mathematics and was yet able to build

[1] "Non conviene tenersi in un semplice equilibrio, ma raddoppiare le resistenze".

the dome." Mathematics is a most respectable science, but in this case it has been abused.[1]

A further objection was that the dome would have collapsed long ago if the calculation of a 3 million lb. deficiency in horizontal resistance were correct. "Heaven forbid that the calculation is correct. For, in that case, not a minute would have passed before the entire structure had collapsed."[2] This objection is not quite without justification. For, the three experts may indeed have failed to point out sufficiently, that the stated value of the thrust represents a *maximum*, based on certain unfavourable, non-realistic assumptions, such as the neglect of the tensile and shearing strength of the masonry, and the assumption of frictionless rotation of the individual parts around rather arbitrarily determined centre points.

But the most serious objections against the "System of the Three Mathematicians" came from *Giovanni Poleni*[3] of Venice who, in 1743, was likewise entrusted with an examination of the cupola. With all due respect to the scientific achievement of the Roman expert committee, Poleni questions as being too theoretical the thesis that the drum segments and buttresses should have carried out a kind of rotation around their base (cf. Fig. 37). Probably with some justification, he prefers some more natural explanations of the damage which had become apparent in the course of time, mentioning external causes such as earthquakes, thunderbolts and the like, and internal causes, including inexpert execution of the masonry. But he

[1] "Se potè la Cupola di S. Pietro idearsi, disegnarsi, lavorarsi senza i Matematici, e nominatamente senza la Meccanica coltivatissima d'oggi giorno, potrà ancora ristorarsi, senza che richieggasi principalmente l'opera de'Matematici, e della Matematica. . . . Buonarroti non sapeva di Matematica, e pur sempre seppe architettare la Cupola. . . . Perchè appunto ho grandissima stima di questa Scienza, altamente me ne dispiace il suo abuso." (Quoted by Poleni).

[2] "Ma Dio guardi che la bisogna fosse andata cosí come i calcoli dimostrano, che non ci voleva neppure un minuto intiero di tempo per far andare la Mole tutta per terra." (Quoted by Poleni).

[3] Giovanni Poleni (1685-1761), mathematician and engineer, was called, at the age of 26, to the University of Padua where he successively occupied the Chairs for Astronomy, Physics, Mathematics and finally, Experimental Philosophy. He was mostly concerned with hydraulics and hydraulic engineering, as a practical engineer as well as a theoretician, scientific author and editor of Frontinus' "De aquæ ductibus".

mainly blames the unequal distribution of the dome weight on the cylindrical drum wall on the one hand, and on the opposing buttresses on the other hand which had, in his opinion, led to unequal settlements and thus to fractures. However, in spite of their difference of opinion on the *causes* of the damage, both Poleni and the Roman mathematicians were agreed on the measures to be taken, namely, the provision of further tie rings to secure the structure.[1]

In accordance with the suggestions of the scientists consulted, five additional tie rings to secure the cupola were built in by the architect, *Vanvitelli* (1700–1773) in 1743/44. It was perhaps partly due to the typically rationalist faith in science, and confidence in the results of theoretical investigations and calculations, that much more far-reaching proposals were rejected, such as the filling-in of the staircases and recesses in the piers, or even the removal of the lantern to reduce the weight, or the erection of four colossal buttresses to support the cupola which had also been suggested.

In spite of individually justified objections, the report of the three Roman mathematicians must be regarded as epochmaking in the history of civil engineering. Its importance lies in the fact that, contrary to all tradition and routine, *the stability survey of a structure has been based, not on empiric rules and statical feeling, but on science and research.* Its importance lies, furthermore, in the new approach to the problem which is, for the first time, treated as a problem of quantitative statics in the modern sense : the dimensions of a building element (the tie ring) are to be determined directly, by calculation. As a matter of principle, this was a vital step forward, and it is of little relevance in this connection that the offered explanation of the damage was too one-sided and artificial, and that the calculation was not correct in all details. The next generation of civil engineers was to come considerably nearer to the goal.

Meanwhile, the polemics between engineering practicians and theoreticians, launched by the report of the three mathematicians, dragged on for more than half a century. At the outset,

---

[1] cf. Poleni, Chapters XIII, XIV, XLVIII, LXIV, LXV and others.

the theoreticians still find it necessary to be quasi-apologetic about the unaccustomed and novel application of mathematics to the purposes of engineering technique. Similar to that of the three mathematicians (cf. footnote 1, page 112) is, for instance, the tenor of a treatise on hydraulic engineering,[1] published in 1741 by *Bernardino Zendrini*, who says in the introduction : "If anyone should perhaps marvel at finding a treatise on watercourses full of algebraic figures. . . . If he will, however, consider the fact that geometry is based on analysis, just as the science of the watercourses is based on geometry . . . he will agree that there can be no method more natural and, incidentally, more concise and safe, than the one used in our treatise."

During the early nineteenth century, however, the theoretical, scientific approach to structural problems is gradually beginning to be taken for granted, and it is the one-sided, anti-scientist practician who finds himself on the defence. Yet, as late as 1805, *C. F. Viel*, Architect of the Paris Hospitals and Member of the Public Works Council, published a treatise entitled "De l'impuissance des mathématiques pour assurer la solidité des bâtiments", which includes, *inter alia*, this sentence : "In the sphere of architecture, in order to probe the solidity of buildings, the complicated calculations, bristling with figures and algebraic quantities, with their 'powers', 'roots', 'exponents' and 'coefficients', are by no means necessary".[2]

Even more malicious is an utterance of the Englishman, *Tredgold* (1788–1829), who had worked his way up from a journeyman carpenter to become a distinguished engineer and who, in his work "Practical Essay on the Strength of Cast Iron and Other Metals" (First Edition, 1822), remarks that "the stability of a building is inversely proportional to the science of the builder".[3] Admittedly, however, the opprobrium of an anti-scientific attitude would be wholly out of place in the case of Tredgold who, indeed, regarded the furtherance of scientific engineering research as a principal aim of his life. The above-

[1] "Leggi e fenomeni, regolazioni ed usi delle acque correnti." Venice, 1741.
[2] According to A. G. Meyer, page 38.
[3] Quoted by F. Stüssi, *Schweizerische Bauzeitung*, Vol. 116, page 201, 2nd November, 1940.

117

mentioned remark may have been caused by the observation that the theoretician of his time overstepped the mark even more often than his modern counterpart, being insufficiently conscious of the limitation of his methods, and of the precariousness of the assumptions on which his calculations were based.

3. "GÉNIE" AND CIVIL ENGINEERS IN FRANCE — ENGINEERING LITERATURE OF THE EIGHTEENTH CENTURY

The term "Engineer" has long been applied[1], in Italy, France and England, to the builders of war machines and fortifications. The word may have originated in the fact that the technical aids of warfare and defence used to be known under the joint term "ingenia". From the fifteenth century onwards, the term "Engineer" is encountered more frequently. In Italy, surveyors and canal builders were sometimes also called "ingeniarii".[2]

The direct ancestors of the modern *civil* engineer, however, were the French "Génie" officers who, apart from their military tasks, were also entrusted with public works of a civilian character (hence the term : "Génie Civil"). It was still under Louis XIV, that *Sébastien le Prestre de Vauban* (1633–1707), famous builder of numerous fortresses, brought the system of polygonal and star-shaped fortifications to a peak of perfection. Most of the capacious walls and moats, modelled to this system and applied to nearly every major city of Central Europe during the eighteenth century, were razed during the nineteenth century. One of the few specimens almost completely preserved is the Alsatian frontier fortress of Neu Breisach (Neuf Brisach), which was built by Vauban himself. A fine view of its walls and moats, profusely covered with growth, can be enjoyed from the regular Zürich–London airliner.

Vauban was, first of all, a soldier. In 1678, he became Inspector-General of the French Fortresses and in 1703,

---

[1] This habit can even be traced to the Middle Ages. Feldhaus ("Zeitschrift des Vereins Deutscher Ingenieure", 1906, page 1599) quotes examples from the twelfth and thirteenth centuries.

[2] cf. Albenga in "L'Ingegnere", 1928, page 548.

"Maréchal de France". As a military man, he took part in more than fifty sieges and more than a hundred engagements. But, in connection with the design of fortified places and major, comprehensive defence systems, he also had frequent opportunities to plan and execute public works which were not only of military significance, but were just as important for peaceful, commercial purposes. In particular, he saw in the canalization of rivers and the construction of inland waterways an efficient instrument to assist the defence of his country in times of war, and to further her commerce and wellbeing in times of peace. During a tour of inspection in Northern France, he conceived the idea of connecting the ports of the Flanders coast with each other, and with their hinterland as far as Lille, Cambrai and Valenciennes, by means of a system of canals and navigable rivers. His was also the idea of a Rhine-Rhone Canal, inherent in the proposal to connect the Alsatian canal system, across the Burgundy Gap, with the river Doubs and thus with the Saône and Rhone.

One of Vauban's most eminent undertakings in the sphere of public works was the conversion of Dunkirk into an impregnable coastal fortress. Apart from the military works proper, i.e. the construction of several forts, there were extensive harbour and coastal works to be carried out : the excavation of canals and harbour basins, the construction of two long jetties flanking the entrance channel, the erection of storehouses and workshops (Fig. 33). A masterpiece of civil engineering was the great lock marking the entrance to the Inner Harbour, which was designed by Vauban himself and built under his personal supervision. The fortress was demolished no more than 30 years after its completion, in consequence of the conclusion, unfortunate for France, of the Spanish War of Succession. Vauban also played an important part in the construction of other French ports of the Atlantic and Mediterranean coasts.

Occasionally, the famous fortress engineer was also called in to advise on works serving exclusively peaceful purposes, such as the water supply system of the Park of Versailles. In this connection, he designed, and partly carried out, the aqueduct of Maintenon-Eure, which was, however, never completed. For a time, he also took part in the completion of the Canal du Languedoc (cf. page 133).

Vauban's projects are masterpieces of engineering methodics and lucidity. They usually consisted of an explanatory memorandum ("Mémoire"), several sheets of drawings, and a covering letter ("Lettre d'envoi"). The "Mémoire" was divided into four sections : (1) General background of the scheme ; (2) detailed description of the constituent parts, with references to the drawings ; (3) cost estimates ; (4) features and advantages of the work.[1]

In addition to numerous works on technical (mainly military) subjects, Vauban also wrote several works on economic questions, including a famous book, printed just before his death and entitled "The King's Tithe". In this work, the author, who had a pronounced social feeling, rare at the time, pleads for a juster system of taxation for the suppressed lower classes.

It was in Vauban's time that the term "Ingénieur" was first used, in France, as a professional title for a scientifically trained technician in the public service, contrary to previous practice, when the word merely served as a vague description of technical practicians without higher education. Vauban himself endeavoured to improve the material and social standing of the Génie officer. On his suggestion, the Minister of War, Louvois, created in 1675 a special organization of military engineers, the "Corps des ingénieurs du Génie militaire", thus partly protecting them against the high-handedness and moods of the generals. In 1689, the naval architects were accorded, through a Royal Decree, the title "Ingénieurs—constructeurs de la Marine". In 1720, the famous "Corps des ingénieurs des ponts et chaussées" was created.[2]

In contrast to non-military builders and architects, most of whom still came, through practical training, from the artisan or artist classes, the French Génie officers enjoyed a scientific education, with special emphasis on mathematics, at State colleges and institutions. The "Ecole des ponts et chaussées",

---

[1] cf. Georges Michel, "Histoire de Vauban" ; Paris, 1879.

[2] cf. Schimank, "Das Wort Ingenieur," in *Zeitschrift des Vereins Deutscher Ingenieure* of 18th March, 1939.

founded by Trudaine in 1747 and reorganized by Perronet in 1760, was unique of its kind in Europe. At this college, a great number of excellent engineers received a training which secured the Continent-wide supremacy of French road and bridge building for a long time to come. The different Génie and Artillery Colleges like those of Mézières (opened in 1749) and La Fère may also be mentioned as the breeding grounds of numerous mathematically trained engineers.

The "Ingénieurs des ponts et chaussées" who, by virtue of their military organization and mathematical schooling, were mainly concerned with public works, were now systematically trying to use the exact methods of mathematics, geometry and statics to determine the dimensions of structures, retaining walls and other important building elements, and to profit from the results of strength tests. Conversely, the practical tasks with which they were confronted induced the engineers to amplify these scientific methods and to adapt them to their special requirements, as well as to deepen the knowledge of material properties through numerous exacting tests.

It soon became necessary for the engineer to be armed with scientific data in a concise and handy form. This need called forth the first specialized manuals and reference books in the field of civil engineering, which contained rules for construction and dimensions, tables of much-used figures and the like, without any literary ambitions and without much historical and anecdotical embellishment. Like our modern reference books, they enjoyed great popularity with the experts, as is apparent from the number of editions which some of them went through. One of the first works of this kind was the "Science des Ingénieurs", first published in Paris in 1729, which was repeatedly re-issued, up to 1830. The author, *Bernard Forest de Bélidor* (1697–1761), was a technical officer and a teacher of mathematics and physics at the Artillery College of La Fère. Under Louis XV, he took part in several campaigns during the years 1742–1745. In 1758, he became Director of the Paris Arsenal and Inspector-General of the technical troops. Apart from the handbook already referred to, Bélidor contributed to the scientific engineering literature a number of books on military and fortress engineering. His "Architecture hydraulique", a work of four volumes on hydraulics and hydraulic engineering, contains, in addition to elementary physics (mechanics, hydraulics, some theorems of the theory of heat, etc.), an exhaustive description of contemporary mechanical engineering. It is, incidentally, the first work destined for technicians, which contains examples of practical applications of the integral calculus. The work will again claim our attention in the following section, which deals with hydraulic engineering.

The "Science des Ingénieurs", which is specifically intended for civil engineers, already shows, in some respects, a certain likeness to our modern "manuals". This applies particularly to the tabular juxtapositions, showing the specific gravities of the most important building materials, etc.,[1] the dimensions of retaining walls and the like, as well as the results of bending tests on wooden beams. Much space is devoted to the subject of tenders and contracts (cf. page 135 *et seq.*). Finally, there are rules and examples for the construction of various military buildings, and a concise manual of architecture, with a textual and pictorial description of the five classical orders of columns.

As far as our subject is concerned, interest is centred on the author's treatment of building statics, and the strength of materials. Bélidor does indeed attempt to deal with the theories of earth pressure, vaulting and bending on a mathematical basis. Although the results are still rather modest, it is nevertheless significant that a practical engineer, writing a work intended for, and much used by his colleagues, does deal with such problems and tries to tackle them by means of exact, scientific methods.

With regard to earth pressure, the author departs from the assertion that it is "a matter proved by experience that normal soil will automatically assume a talus of 45 degrees".[2] He is therefore dealing with an earth prism which is represented by a right-angled, isosceles triangle, and which slides on a

---

[1] The new tabular form of presentation solicited comprehensiveness : a table of specific gravities of liquids, contained in Bélidor's "Architecture hydraulique", shows also the weights of cow's milk, goat's milk, asses' milk as well as of champagne and red Burgundy wine !

[2] "Une chose démontrée par l'expérience que les terres ordinaires . . . . prennent d'elles-mêmes une pente ou talud . . . de 45 degrés" (Bélidor, "Science des Ingénieurs", Book I, Chapter 4).

plane inclined at 45°. According to the rules of the inclined plane, and ignoring friction, the horizontal pressure of that earth prism on a vertical retaining wall is equal to its dead weight. But as Bélidor, the engineer of the practice, regards this value as too high, he suggests that the actual horizontal thrust should be assessed at half that value, having regard to the ever present friction. The stability of the retaining wall under the load of the earth pressure is examined on the basis of the lever principle, taking the bottom front edge of the wall as the axis of rotation. In relation to that axis, the stability moment of the dead weight must be greater than the tilting moment of the earth pressure.

Just as inadequate as Bélidor's calculation of the earth pressure is his treatment of the theory of vaulting, which is based

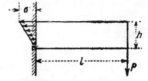

Fig. 38. Flexure, according to Leibniz and Bélidor.
Reproduced from *Schweizerische Bauzeitung*, Vol. 116.

on the research of De la Hire (cf. page 139 *et seq.*). Having discussed the equilibrium of the individual voussoirs as determined by their weight and by the reaction of the adjacent voussoirs, Bélidor again refers to the friction in the mortar joints to explain the obvious stability of the semicircular arch observed. The presentation is, however, rather confused, and it seems that Bélidor had not yet a really clear idea of the action of forces in a vault. To determine the dimensions of the supports, he follows a similar trend of thought as in the case of retaining walls ; but he also quotes the old rule of Blondel (cf. page 92).

Dealing with the theory of bending, the author mentions Parent ; but, although he follows the example of Leibniz and Varignon in assuming a triangular distribution of the tensile stresses, he places the neutral axis again at the lower edge of the cross-section and thus obtains a wrong value for the relation of bending and tensile strength ($W = \dfrac{bh^2}{3}$; cf. Fig. 38).

But, in spite of such individual shortcomings, Bélidor's systematic and, at the same time, practical approach to the subject matter represented a great step forward towards modern engineering science. This is, for instance, apparent from a comparison with another popular book, published in Italy in 1748, i.e. nearly twenty years later, under the title : "Istruzioni pratiche per l'ingegnere civile". This book, written by *Giuseppe Antonio Alberti* (1712–1768), went through nine editions (the last of them published in 1840). The first part of this publication deals with surveying and explains the topographical instruments and their application, especially the optical distance gauge used for the plan table method. The second part, devoted to building practice, is concerned with subsoil works, earth movements, river regulation, and the construction of dams and weirs, bridges and aqueducts, locks and harbours. Here, the treatment of the subject matter is merely descriptive. There is no mention of the theory of statics or that of the strength of materials, and one looks in vain for any indication of the relatively high standard which the mechanical sciences had, after all, already attained at that time. The whole is more like a catalogue for artisans and artists, although there is, true to humanistic pattern, a great show of historical knowledge. In the different chapters on bridges, canals, harbours, the most important works of Antiquity are thus listed and briefly described, naming those by whom the work was commissioned, and the relevant literary sources. To mention but one example, the author tries, in the chapter on bridge building, to prove the possibility of bridging even the greatest rivers by referring to Queen Semiramis' bridge across the Euphrates, Cæsar's bridge over the Rhine, and Trajan's bridge over the Danube.[1]

Similarly, the contemporary civil engineering literature in the German language hardly rises above the level of an artisans' prospectus, without scientific character. In the sphere of bridge

---

[1] "Possonsi far Ponti sopra qualsivoglia fiume, benchè vasto e rapido, che che se ne dica in contrario. Abbiamo l'esempio in Diodoro Siculo che racconta che Semiramide, Regina degli Assirj, fece fare un ponte sopra l'Eufrate . . . che era lungo cinque stadj . . ." etc. (Part II, Chapter X : "De Ponti Sopra i Fiumi, e modo di construirli").

building, it is the wooden bridge which, in accordance with local tradition, is treated in greater detail, thus in a booklet on carpentry written by Johann Wilhelm,[1] in L. C. Sturm's "Architectura civili-militaris" (1719) and in Lukas Voch's "Brückenbaukunst" (1780).

Among the French public-service "Ingénieurs des ponts et chaussées" with scientific background is also *Jean Rodolphe Perronet* (1708–1794), famous as the builder of many classic masonry bridges, including the Pont de Neuilly and Pont de la Concorde over the Seine in Paris. The former, built in 1768–1774, survived until 1939. The latter, erected in 1787–1791, was enlarged in 1930–1931, but the façades with the beautifully cut stone have been preserved.

Perronet came from Château-d'Oex in Switzerland, but was born in France as a son of a Swiss officer in the Royal service. He rendered a signal service to French civil engineering through the complete reorganization of the "Ecole des ponts et chaussées", already referred to. The school originated in the official "Bureau central des déssinateurs des ponts et chaussées" which Perronet took over in 1747 after some road engineering practice. Since 1750 Chief Inspector and later (1763) Chief Engineer of the "Corps des Ingénieurs des ponts et chaussées", he was mainly concerned with bridge building, where he introduced important innovations. In order to enlarge the flow clearance, he reduced the width of the piers below the dimensions prescribed by conventional rules,[2] and resorted to flattened polycentric or segmental arches instead of the conventional, more or less semi-circular arches. The same purpose of easing the passage of the floods is pursued in his design of the so-called "horn" (oblique flattening of the edges of the arch in the vicinity of the abutments) which were, for instance, apparent on the Pont de Neuilly. Carefully prepared foundations and a

---

[1] "Architectura Civilis oder Beschreibung und Vorreissung vieler vornehmer Dachwerk als hoher Helmen, Kreutzdächer . . . Fallbrücken . . . item allerley Pressen, Schnecken . . . und andern dergleichen Mechanischen Fabricken" ; Nuremberg, 1668.

[2] "La connoissance que nous avons de la force des pierres pour résister au poids . . . nous a fait penser qu'on pourroit diminuer de beaucoup l'épaisseur qu'on est dans l'usage de donner aux piles, laquelle est évaluée ordinairement au cinquième de l'ouverture des arches" (Perronet, see Bibliography).

skilful bond guaranteed the safety and durability of his structures in spite of their boldness.

Perronet was also active in the field of hydraulic engineering, one of his works being the Canal de Bourgogne. As a technical author, he has described the designs and construction methods of a number of his bridges, including those at Neuilly, Mantes, Orléans, etc., in an extensive volume[1] of large size, with numerous magnificent engravings. These show not only the structures themselves during the different phases of construction but also, in great detail, the installations, devices and machines used. The work affords an interesting insight into the state of contemporary building technique and shows, incidentally, how the bridge builders, prior to the invention and introduction of heat engines, were able to utilize the energy of flowing water for the operation of building plant and machinery, such as the bucket wheels used for the drainage of foundation pits. On one occasion, for the foundation of the bridge of Sainte Maixence, Perronet goes as far as to use a water wheel, worked by the river itself, for the lifting of the 2,000 lb. rammer of the pile driver (cf. Figs. 34, 35 and 39).

Among other works on bridge building, published by Perronet, are "Sur les pieux et sur les pilots ou pilotis", "Sur le cintrement et le décintrement des ponts", "Sur la réduction de l'épaisseur des piles". Without actually submitting any difficult, statical calculations, the author seeks to persuade his colleagues to adopt a scientific method of approach, and to utilize the results of research, especially those of strength tests, for engineering purposes.

His essay on pile foundations contains, apart from practical rules on the length, thickness, spacing and quality of the piles, some particulars regarding their driving resistances. The piles should be driven until the penetration, during the last "heat" of 25–30 blows, does not exceed 1–2 "Paris Lines",[2] or 6 Lines in the case of less loaded piles. The driving force of the rammer is proportional to the height of its fall, but "On n'ignore pas combien il est difficile . . .d'établir mathématiquement aucun

---

[1] "Déscription des projets et de la construction des ponts de Neuilli, de Mantes, d'Orléans . . . etc."

[2] 1 Line=1/12 inch.

rapport entre les forces mortes (static forces) et les forces vives".

His treatise on centring and decentring contains, *inter alia*, advice on the vaulting process ; on the camber to be applied to the centring in order to compensate for the settling of the arch ; and, most important of all, on the striking of the centres. In contrast to the traditional practice of removing the centring progressively from the abutments to the crown, his recommendation is to lower the *entire* centring simultaneously, at a slow and uniform rate.

In the last of the three publications named, Perronet expresses the opinion that an exaggerated thickness of the piers is not only superfluous but even harmful : the narrowing down of the flow clearance will increase the flow velocity and, thus, the risk of cavation which has already been responsible for the collapse of many a bridge.

One of Perronet's pupils was *Emiland Marie Gauthey* (1732–1806), a student and, for a time, teacher at the "Ecole des ponts et chaussées". He was another of France's distinguished engineers who made his career in the public service, first as Engineer-in-Chief of the Province of Burgundy, then as Director of the Burgundy Waterways and, finally, as "Inspecteur-Général des Ponts et Chaussées" in Paris. He continued the tradition of his great teacher, not only as a builder of numerous bridges, but also as a theoretician and technical writer, although his main work, "Traité de la construction des ponts", was published only after his death, by his nephew, Navier.

As a technical adviser to Soufflot and Rondelet, Gauthey was also concerned with the building of S. Geneviève's Church (Panthéon) in Paris. The construction of this edifice was commenced in 1757, but dragged on for several decades. Already before the vaulting of the cupola, there had been differences of opinion on the question of whether the piers, which were slender compared with similar buildings, would be able to resist the weight and thrust of the dome. In a treatise published in 1771 and entitled "Mémoire sur l'application de la mécanique à la construction des voûtes et des dômes", Gauthey sought to allay Patte's misgivings and to prove that the thrust of cupolas is generally over-estimated. Gauthey uses, in his calculations, the method recommended by De la Hire (see the following chapter, Section 1) and arrives at the conclusion that, with the dimensions envisaged, the stability would, in fact, be guaranteed.

Nevertheless, when the centring was removed after the completion of the cupola, there were early signs of serious damage which was, however, found to be due, not to inadequate dimensions, but to negligent execution. Detailed examinations were carried out which proved the need for thorough restoration works. Gauthey and Rondelet, who belonged to the Building and Supervisory Committee, approached the task with exact-scientific methods. For this purpose, they carried out detailed calculations as well as extensive strength tests (already referred to on page 110), which Rondelet has described in detail in his "Mémoire historique sur le dôme du Panthéon" and in his monumental work, "L'art de bâtir". The last-named book is, incidentally, also notable for the fact that it is the last universal work, before the impending specialization, to combine all aspects of building construction — art as well as statics, architecture as well as engineering.

As far as the development of structural statics is concerned, the most important of the French Génie officers was Coulomb, to whom a special section will be devoted (page 146 *et seq.*).

4. HYDRAULICS AND HYDRAULIC ENGINEERING IN THE SEVENTEENTH AND EIGHTEENTH CENTURIES

Long before building construction split into structural engineering and architecture, the construction of dams, canals, irrigation systems, locks and the like had formed a special branch of civil engineering. The artists and builders concerned with the design and execution of hydraulic works needed a knowledge of geometry and mechanics and had to be familiar with civil engineering practice proper. But like the theories of statics and strength of materials, the science of hydraulics owes its origin to physicists, and not to engineers. After Archimedes (cf. page 24), the essential principles of hydrostatics were first restated during the Renaissance, by Stevin, Galilei[1] and others. Galilei's pupil *Torricelli* (1608–1647) laid the foundation of the theory of flow when he found, in 1643, that the mean velocities of a liquid flowing out of the bottom outlet of a vessel are proportional to the square root of the pressure head (i.e. the column of liquid above the outlet). The analogy with the fall of solid bodies may have aided Torricelli in formulating this law. Subsequently, *Mariotte*[2] determined, experimentally, the flow volume for different outlets and found that, with an opening of about ¼ in., the volume actually measured was no more than about 7/10 of the theoretical figure. In 1695, *Varignon* attempted to find a theoretical proof for Torricelli's theorem.

As in the sphere of solid mechanics, the first attempts to apply the scientific discoveries of hydraulics systematically to practical tasks were made by the French engineers of the eighteenth century. *Chézy*[3] (1718–1798) laid the foundations of the theory of uniform flow in a river or canal bed when he discovered that a relation similar to Torricelli's law existed between the mean velocity of flow (u), the fall (J) and the mean depth of the water course. Written in its modern form, the formula expressing this relation which is still named after the French engineer, is $u = c \sqrt{RJ}$. Herein is 'c' a constant dependent on the local circumstances, and R the so-called "hydraulic radius", which is nowadays used instead of the mean depth, and which represents the cross-sectional area of the water course, divided by the wetted perimeter.

The notion of the "hydraulic radius" was introduced by *Du Buat* (1734–1809), who was Chézy's junior by sixteen years. This "Capitaine d'infanterie ingénieur du Roi" concerned himself, in great detail, with engineering hydraulics, i.e. with the flow of water in rivers, canals and pipelines. He endeavoured to reconcile the actual flow phenomena with the principles of mechanics. For this purpose, he carried out numerous experiments which made him one of the foremost experimental

---

[1] "Discorso intorno alle cose che stanno in su l'acqua."
[2] "Traité du mouvement des eaux."
[3] Ingénieur des ponts et chaussées ; a theoretician who served as an assistant to Perronet and became the teacher of Prony.

research workers of his time.[1] But the time was not yet ripe for the birth of scientific engineering hydraulics (cf. Chapter VIII, Section 1).

Bélidor's "Architecture hydraulique" (cf. pages 121–22) combines, in one great work, the subjects of mathematical physics, mechanical engineering and hydraulic engineering. The first volume deals with the principles of mechanics, with special reference to their application to elementary machines, as well as with the foundations of hydraulics in as far as they were known at the time : hydrostatics, capillarity, flow of water out of a vessel. In this connection, the discrepancy between the actual mean flow velocity and the greater theoretical value, $\sqrt{2gh}$, is ascribed to the influence of friction. The second volume contains, *inter alia*, a chapter on the theory and construction of suction and pressure pumps, with numerous examples of actual installations. One chapter is devoted to the "machines à feu" as Papin's steam engines, exclusively used for the operation of pumps, were then called. Another chapter deals with distribution pipes, and a last chapter with ornamental fountains (Versailles). The third and fourth volumes are concerned with hydraulic engineering as a branch of civil engineering, including the design and construction of ship locks, harbour installations and dry docks, river regulation and canal construction, land drainage and irrigation works. The author reveals a profound knowledge of deep level construction, and of the plant required for it. Of special interest are the machines for under-water works, such as the digger used at Toulon for the excavation of muddy ground (Fig. 40), or the enclosed boxes with drop bottom used for the pouring of under-water concrete.

With the construction of the Atlantic and Channel Ports which are particularly exposed to the tides, the most important engineering works were the *locks*. To overcome the level differences of inland waterways, such installations were already in use since the Renaissance. Leonardo da Vinci had not, in fact, invented them as is sometimes alleged, but he may have perfected them through the development of an improved

---

[1] cf. R. Dugas, "Histoire de la Mécanique" ; Neuchâtel, 1950, page 303.

Fig. 39. Construction of the centering for the Pont de Neuilly

From Perronet, *Déscription des projets* ...

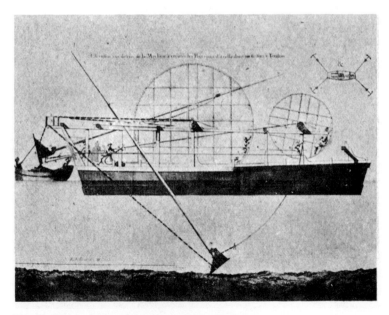

Fig. 40. Dipper dredger for under-water excavation

From Bélidor, *Architecture hydraulique*

Fig. 41. Ship canal   *Vignette from* Bélidor, *Architecture hydraulique*

Fig. 42. Canal du Languedoc, lock near Carcassone   *Photo by courtesy of*
Bernon, Carcassone, Société des Auteurs photographes du Languedoc

gate design. Applied to the tidal ports, the locks made it possible to retain, also during the low tide, the depth of water required for the sea-going vessels anchored in the inner basin. The French engineers of the eighteenth century also used the locks as a means of deepening the outer port and approach channels, through opening the locks at low tide and thus releasing the water of the inner port at high speed. "These locks had an amazing effect which could not have been achieved by a multitude of men in a long time."[1] Already during the seventeenth and eighteenth centuries, locks were in existence at Muiden (Holland), Dunkirk, Gravelines, Calais, Le Hâvre, Cherbourg and other places ; some of them were of considerable dimensions, with several chambers or two parallel passages.

Problems similar to those encountered with locks confronted the harbour engineer during the construction of *dry docks* for the repair of major sea-going vessels. The oldest installations of this kind were developed from slightly inclined slipways, by enclosing their upper part with wooden or stone walls and equipping these with a wooden lock gate which could be closed at high tide. The vessels were able to enter during the high tide only.

At Portsmouth, a dock of this kind is already mentioned in 1496, though without lock gate, the entrance being closed by a kind of cofferdam of clay and stone. In 1666, an order was placed for a new installation, at a cost of £2,100. The larger of the two dry docks, built in 1703 at the Howland Wet Dock on the Thames, measured 247 ft. in length, 44 ft. in width and $16\frac{1}{2}$ ft. in depth during the high tide, and was able to accommodate the largest merchantmen of the time.[2]

In his "Architecture hydraulique", Bélidor describes several dry docks in France, some of which were already equipped with mechanical pumping installations, so-called "machines" (e.g. at the Mediterranean ports where the tidal rise is small). His descriptions include one such dock at Marseilles, a two-

---

[1] "Elles faisoient un travail prodigieux, que n'auroit pu exécuter en beaucoup de tems une multitude infinie d'hommes," Bélidor, "Architecture hydraulique", Vol. III, page 28. The volume also contains drawings and descriptions of actual installations.

[2] cf. H. Ridehalgh, *Dock and Harbour Authority*, November 1947, page 174.

chamber dock at Dunkirk, and two docks at Brest which were still under construction at the time.

The sealing of the bottom, and of those parts of the side walls which were lying below the low-water level, confronted the engineers of the time with almost insuperable difficulties. In spite of mechanical water drainage, they succeeded only with the greatest effort in keeping the dock chamber dry during the repair of a vessel. Bélidor therefore recommends the sole to be placed as high as possible and, if necessary, to resort to a two-chamber installation where the ship is raised to the interior high-level chamber by means of pumped-up water.

An epoch-making event, especially due to the novel construction method, was the building, in 1774, of a large dry dock in the harbour of Toulon by *Groignard*. For this purpose, the French engineer built a large wooden floating caisson of 31 × 100 metres base and 11 metres height which was partitioned by cross bulkheads into eight compartments. The ground having been levelled beforehand, the floating caisson was sunk, by means of stone ballast, at the site selected for the dry dock. Although the walls had been carefully sealed, it was only with the greatest effort that the interior could be laid dry. This being done, sole and side walls were built, and the work completed, after a great many further difficulties.[1]

Apart from harbour construction, the construction of *ship canals* was the most important task of the public works engineers of the seventeenth and eighteenth centuries. In spite of Colbert's efforts in the sphere of road construction, the highways of the then most populous country of Europe, proved to be more and more inadequate for the increased needs of transportation. The water-borne conveyance of bulk goods was therefore regarded as an important step forward. Already at the time of Henry IV, Sully began the construction of a canal connecting the rivers Seine and Loire where, presumably for the first time in France, locks were built to overcome the difference in level.[2] The reign of Louis XIV and his Minister, Colbert, saw

[1] cf. P. Bonato, in *Annali degli Ingegnere ed Architetti italiani*, Rome, 1888, No. 2.
[2] Bélidor, "Architecture hydraulique", Vol. IV.

the construction, in 1667–1681, of the famous Canal du Langue-doc or Canal du Midi which connects the Mediterranean port of Sète with the Garonne near Toulouse, and thus with the Atlantic Ocean (Figs. 41 and 42). The canal, which is still in operation in its original form, was planned by *Riquet* (1604–1680) who was also mostly responsible for the supervision of the construction work. About 150 miles long, the canal rises, by means of 74 locks, to an altitude of about 620 ft. above sea level and descends to the Garonne by means of another 26 locks.[1] Several high-level storage reservoirs serve to feed the canal during the dry season. Having a width of about 33 ft. and a draught of about $6\frac{1}{2}$ ft., the canal can carry barges of 200 tons.

The construction excited the admiration of the contemporary world. "There is, in fact, nothing in the world more deserving of admiration than the sight of ships passing from one sea to the other, 600 ft. above their port of departure."[2] Voltaire, in his classic "Siècle de Louis XIV," having mentioned the Louvre, Versailles, Trianon, etc., concludes the chapter dealing with the King's building activities with these words : "Mais le monument le plus glorieux par son utilité, par sa grandeur et par ses difficultés, fut ce canal du Languedoc qui joint les deux mers. . . ."

Subsequently, numerous other ship canals were built in France. Most of the "Ingénieurs des ponts et chaussées" were also concerned with the construction and maintenance of waterways. Perronet built, as already mentioned, the Canal de Bourgogne which links the Yonne and Saône in the vicinity of Dijon. Gauthey was responsible, in his capacity of the Director of Waterways in the Province of Burgundy, for the Canal du Centre between Chalon sur Saône and the Loire, and for the Saône-Doubs Canal.

In England, too, ship canals were built during the eighteenth century. The facilities thus created for easier transport of larger quantities of bulky raw materials were one of the prerequisites

---

[1] Bélidor, "Architecture hydraulique," Vol. IV.

[2] "En effet, y a-t-il rien dans le monde de plus digne d'admiration que de voir des bâtimens passer d'une mer à l'autre, en parcourant une partie du pays, élevés de six cens pieds au-dessus du port d'où ils sont partis." (Bélidor).

of the industrial revolution which commenced during the reign of George III (1760–1820). England's most distinguished canal engineer was *James Brindley* (1716–1772) who, unlike the French State engineers, had received practically no school education and acquired his profound technical and mechanical knowledge solely through engineering practice. Canals of more than 365 miles aggregate length have been either designed by him or built under his supervision, including the Grand Trunk Canal between the Trent and the Mersey, and the so-called Duke's Canal between Worsley and Manchester.[1] Collaborating

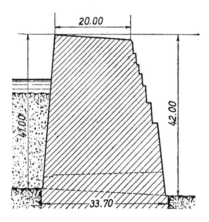

Fig. 43. Dam at Alicante, cross-section. Dimensions in metres.

with Brindley on a number of canal projects was *John Smeaton* (1724–1792), the builder of the famous lighthouse on the Eddystone Rocks and designer of large "fire engines" (atmospheric steam engines) for pit drainage purposes.

Finally, mention must be made of *dams* for irrigation purposes. In this sphere, it was Spain who, prompted by her inveterate water shortage, led the development from the end of the sixteenth century right up to the eighteenth century. The 135 ft. high dam of Alicante (Fig 43), built as early as 1580, remained the boldest specimen of its kind up to the middle of the nineteenth century. (The dam of Puentes, built

---

[1] C. Matschoss, "Männer der Technik". Berlin, 1925.

in 1790, which was some 33 ft. higher, gave way only eleven years after its construction, the wooden pile foundation having been undermined by the water. More than 600 people were drowned in the flood).

In France, the reservoirs needed to feed the ship canals called for the construction of dams. The Canal du Midi has several of them, including two dating back to the construction of the canal itself, i.e. the last third of the seventeenth century. These are the earth dam of Saint-Féréol which is 105 ft. high and sealed by three core walls of masonry, and the dam of Orviel. Another dam, that of Lampy, was built in 1780 when it was found necessary to improve the water supply so as to maintain the operation of the canal during the dry season. This dam (Fig. 47) is of the gravity type, consisting of quarry stone masonry with cut stone facing and reinforcing buttresses.

The Spanish dams were built of quarry stone or cut stone and usually had a rather uneconomic, nearly trapezoidal cross-section with a broad crest (Fig. 43). Most of them were of the gravity type. The dam of Almendralejo near Badajoz which is still in operation today is, however, reinforced by five buttresses on the downstream side.[1] The guiding principles followed by the Spanish engineers of the sixteenth and seventeenth centuries in determining the dimensions of their dams are no longer known. In France, the dimensions of dams subjected to one-sided water pressure were determined by the tilting moment in relation to their downstream base. In his "Architecture hydraulique" (Vol. III, page 80), Bélidor gives an example of such a calculation, with a simple formula applicable to a rectangular cross-section. He recommends a safety factor of 1.5 but ignores the upthrust in the foundation joint.

5. ENGINEERING SPECIFICATIONS IN THE EIGHTEENTH CENTURY

With the advent of the civil engineer in the modern sense, the system of tendering and contracting also begins to assume

---

[1] cf. "Revista de Obras Publicas." Madrid, 1st June, 1936 ; reviewed in *Annali dei Lavori Pubblici*, 1936, page 549.

forms which are no longer much different from modern methods. The rationalist, scientific mentality of the French military engineers, combined with the spirit of mercantilism pervading the France of Louis XIV and Colbert, also governed the relations between the State and its contractors. Vauban (cf. page 118) was one of the first to lay down standard rules for tenders and contracts which contain precise stipulations regarding the execution of the individual works, origin and quality of the building materials to be used, duties and responsibilities of the contractors, book keeping and accountancy, etc. Many of these stipulations are hardly different from those commonly applied today. Others, however, are typical for the mentality of pre-revolutionary France, and appear incomprehensible to our socially minded age. There is, for instance, a stipulation which lays down that negligent work is not only a financial responsibility on the part of the contractor but also a criminal offence on the part of the individual workmen : "Si quelqu'un de ces Maçons étoit surpris dans son travail faisant quelque maçonnerie à sec et sans mortier, il sera chassé de l'ouvrage et châtie de la prison, et l'Entrepreneur condamné en cent livres d'amende."[1]

In his "Science des ingénieurs", Bélidor devotes much space to tenders and contracts and reflects on the subject in terms which might have been written today : ". . . If it is possible to find contractors, solvent and capable to take on a general contract ('enterprise générale'), one will do well to deal with them. But it is rare to encounter men of strong enough character ('têtes assez fortes') to take on a burden as heavy as that of a general contract. For, the haste with which these works are usually undertaken, and their long duration, often reduce the contractor to a state of nervous exhaustion ('reduisent souvent l'Entrepreneur à ne savoir plus où il en est'). . . . They must be accorded reasonable terms, without pressing for excessive rebates. For, if the job is somewhat heavy, and if it is awarded to poor or ignorant people, they will take it on rashly at any

---

[1] Article 86 of "Devis des ouvrages de terre, charpenterie, maçonnerie, ferrurerie, plomb et bronze, qui ont servi à la fabrique de la grande écluse du bassin de Dunkerque, construite en l'année 1685", quoted by Bélidor, "Architecture hydraulique", Vol. III.

price, in the hope of making some profit in one way or other. . . . The workers, being poorly paid, will desert, and will only turn up in small numbers. All this will give the engineers plenty of headaches. . . ."[1]

The procedure governing the award of public works contracts, in eighteenth century France, to the most favourable bidder is not without a certain picturesque attraction : "On the appointed day, the Superintendent . . . reads out aloud the entire Specifications . . . and the conditions to which they must conform. This is followed by the different tenders of the contractors. Then, if there are no further persons to submit an offer, three candles are lit successively. Whilst these are burning, it is still possible for another contractor to submit a new tender. . . ."[2] A similar procedure, also with a burning candle, is encountered in an Italian regulation which still applied during the present century.[3]

Perronet, too, was not only concerned with the technical and æsthetic aspect of bridge building ; he also paid due attention to the relations with contractors. He regarded the Specifications drawn up by him or under his supervision as so important and exemplary that he reproduced them, in extenso, in his great work on bridge building (cf. page 126). The Specifications for the Pont de Neuilly, for example, contained 260 Articles, and those for the Pont de la Concorde 217 Articles, with the detailed description of the bridges ; precise stipulations regarding the quality, origin and treatment of the building materials to be used ; the construction programme ; site organization ; legal and financial clauses, etc. The rather self-opinionated concluding sentence reflects the important standing of the supervisory engineer : "Le présent devis fait par nous, chevalier de l'ordre du roi, son architecte et premier ingénieur des pont et chaussées . . . signé Perronet."[4]

The relationship between supervisor and contractor, as reflected in the contemporary engineering literature, shows

[1] Bélidor, "Science des ingénieurs", Book III, Chapter 7.
[2] Ibidem, Book VI.
[3] Article 74 of the "Regolamento per l'amministrazione del patrimonio e contabilità generale dello Stato" (R.D.L. No. 827, 23rd May, 1924).
[4] Perronet "Description des projets et de la construction des ponts de Neuilli, de Mantes, d'Orléans, etc." ; Paris, 1788, page 324.

perhaps even more clearly than the beginnings of structural analysis to what extent the French Génie officers and state-employed civil engineers of the late seventeenth and the eighteenth century can already be regarded as civil engineers in the modern sense.

CHAPTER VI

THE ORIGINS OF STRUCTURAL ANALYSIS
IN FRANCE (1750–1850)

I. VAULTS AND DOMES

It was towards the end of the seventeenth century that the first theoretical investigations were carried out on the statical behaviour of vaults which are, after all, among the most important elements of engineering structures. True, the Roman engineers and, especially, the builders of the bold Gothic vaults and buttresses had no doubt a very clear conception of the effect of the arch. They knew that, where one stone block supports its neighbour, it is possible, by means of a skilful bond of wedge-shaped voussoirs, to bridge a gap which is many times greater than the size of the individual voussoirs. During the Renaissance, Leonardo da Vinci and Bernardino Baldi had made occasional attempts to analyze the cause of the arch thrust.

But, apart from isolated attempts (Blondel's and Carlo Fontana's "rules", cf. page 91, were merely empiric and have nothing to do with a theoretical, statical approach), it was *De la Hire* (1640–1718) and *Parent* (cf. page 70) who were the first physicists to investigate the equilibrium of a vault as a mathematical problem of statics. On the assumption that there is no friction at the joints, they analyzed the equilibrium of the individual voussoirs with the aid of the rules governing the composition of forces. In his "Traité de mécanique", De la Hire argues that the shape of the vault must be such that, for each voussoir, the resultant of the dead weight and of the pressure of the preceding voussoir is normal to the face of the next voussoir. In this case, the stability of the structure would be safeguarded even if there were no friction at the joints so that the individual voussoirs could move "comme des corps de surface infiniment polie". This discovery is summed up by the Scottish mathematician, *D. Gregory*, in his treatise "Properties

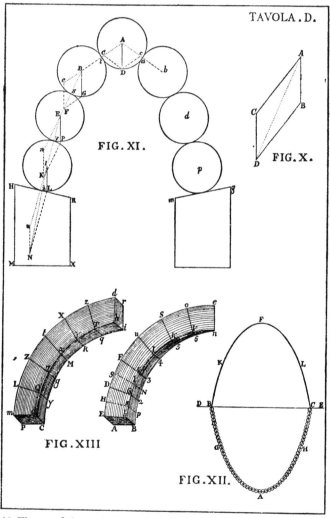

Fig. 44. Theory of the vault thrust line. Balls arranged in accordance with the thrust line.

From Poleni, *Memorie istoriche della Gran Cupola del Tempio Vaticano*, 1748.

of the Catenaria" (1697), in the theorem that the theoretically correct central line of the arch must be shaped like an inverted catenary.[1]

The thesis of the absence of friction can be demonstrated most clearly if the wedge-shaped voussoirs are thought to be replaced by spheres which, being arranged exactly in accordance with the line of thrust, support each other and must thus remain in an unstable equilibrium. Following the example of the Englishman *J. Stirling* (1717), the Italian mathematician and engineer, *Poleni* (cf. page 115) used this method of illustrating the effect of a vault in his report on S. Peter's dome, published in 1748. In doing so, he refers to Newton and deliberately adapts the latter's theorem of the forces parallelogram, originally formulated for dynamic forces, to the conditions of statics (Fig. 44). Poleni's reflections also lead him to regard the line of thrust as an inverted catenary.[2] On this assumption, he determines the correct shape of arches and domes by finding, experimentally, the shape assumed by a chain carrying unequal weights, proportional to the sections of a vault or dome segment (Fig. 45). Small deviations of the actual arch line from the theoretical line are of no significance. All that matters is that the line of thrust does not fall outside the masonry at any point.[3]

Before the invention of concrete, the *stone cut* used to play an important part in solid construction. During the eighteenth century, when the construction of vaults became the concern of mathematically trained engineers, the theory of the stone cut was developed into a proper science. About the middle of the century, *Frézier* (1682–1773) published a work of three volumes on the subject.[4] Having first discussed the elements of two-dimensional and three-dimensional geometry,

---

[1] cf. Hollister, "Three Centuries of Structural Analysis" in *Civil Engineering*, Vol. 8, page 822 (December, 1938).

[2] cf. Poleni, Column 33. A similar presentation of the theory of the line of thrust is also found with Rondelet.

[3] "Che dentro alla solidità della volta la nostra catenaria tutta intiera sia situata" (Poleni, Column 50).

[4] "La théorie et la pratique de la coupe des pierres et des bois pour la construction des voûtes et autres parties des Bâtimens Civils et Militaires", new edition. Paris 1739/54/68.

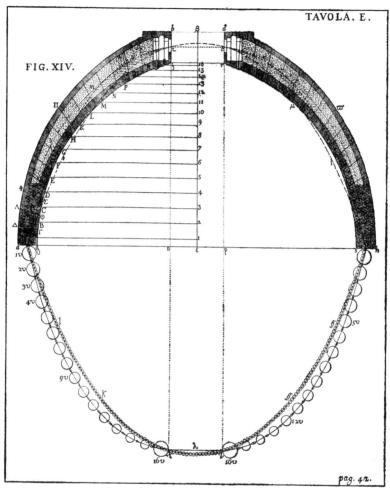

Fig. 45. St. Peter's Dome, Rome. Construction of the thrust line as an inverted catenary.
From Poleni, *Memorie istoriche della Gran Cupola del Tempio Vaticano*, 1748.

the author deals with the construction and graphic presentation of all kinds of vaults, including the most complicated forms of oblique, double-curved and spiral-shaped vaults. Problems like the penetration of cylindrical, conical and spherical surfaces, and the construction of the corresponding intersecting curves,

142

are solved in accordance with the methods of descriptive geometry. Frézier rejects as ugly the theoretical determination of the shape of a barrel vault on the basis of an inverted catenary. But he suggests that this curve, which he acknowledges to be statically correct, can in practice easily be included in the actual thickness of the vault, especially in the case of a pointed arch.

As the failure of vaults is mostly caused by the spreading of the abutments, the calculation of the arch thrust and, in consequence, the determination of the *abutment* dimensions are, if anything, more important to the designer than the exact knowledge of the action of forces in the arch itself.

Independent of mathematical deliberations, practical experiments on models were carried out by *Danisy*, as early as 1732, to determine the permissible minimum thickness of abutments and thus to check the conventional "rules" of the time.

As regards the theoretical treatment of the question, the attempt of the three Roman mathematicians (already referred to in Chapter V, Section 2) to solve the problem by a kind of energy equation was not followed up. As far as the engineers took an interest in scientific calculations at all, they generally followed the method proposed by *De la Hire* which is, incidentally, based on similar statical reflections. In dealing with the problem, the French physicist bases his theory on the alleged observation that vaults with weak abutments usually fail at the quarter points of the arch. He therefore regards the central part of the arch, between the two quarter points, as a wedge which, through its weight, seeks to separate the two lateral parts, which form entities with the abutments. In the process, the monolithical character of the central part of the arch is safeguarded by the adhesion of the mortar. "The need to use mortar makes it unnecessary to calculate the thrust of all the voussoirs individually ; it is sufficient to consider a certain number of them as if they formed a single voussoir only, in order to avoid an excessive length of calculation."[1] Following the rules of the

---

[1] "La nécessité de se servir de mortier fait qu'on peut se dispenser de calculer la poussée de tous les voussoirs, chacun en particulier ; il suffit d'en considérer une certaine quantité, comme faisant ensemble qu'un seul voussoir, afin de'éviter l'extrême longueur des calculs" (Bélidor, "Science des ingénieurs", Book II, page 10).

forces parallelogram, he obtains (ignoring friction in the joint at the assumed point of failure) the resultant of the horizontal thrust and the weight of the arch between crown and quarter point, as a line inclined at about 45 degrees. Stability is secured, if the "stability moment", i.e., the moment of the dead weight of the abutment (including the quarter arch connected with it) in relation to the outer base point, is greater than the tilting moment of the inclined vault thrust (cf. Fig. 46). Gauthey

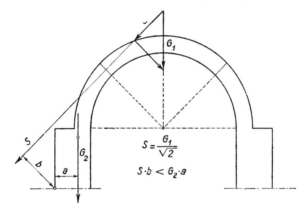

$$S = \frac{G_1}{\sqrt{2}}$$

$$S \cdot b < G_2 \cdot a$$

Fig. 46. Stability survey of a vault, according to De la Hire.

criticizes this theory in his writings on the ground that several important circumstances (friction, adhesion of the mortar in the joints at the assumed points of failure) are left out of consideration. Nevertheless, he himself makes use of this theory in his statical calculations concerning the stability of the Panthéon dome in Paris (cf. Gauthey's "Mémoire", mentioned on page 127).

In contrast to the theory of *flexure* in which the physicists, since Mariotte and Leibniz, tried to take into account the *elasticity* of the material, the stability of vaults was merely regarded as a problem of equilibrium of rigid bodies. True, Gauthey is aware of the inadequacy of this approach. "If a vault were composed of incompressible materials . . . and if there were absolutely no settlement, its stability would be assured, if the abutments had an appropriate thickness, and if the height of the crown were sufficiently great for the stone not to be crushed."[1] But there was still an unbridgable gap between the recognition of the influence, exerted by the compressibility and elasticity of the building material, and the possibility of including these properties in the calculation. Even Coulomb and Navier did not advance beyond this purely *statical* approach to the problem of the fixed arch ; this in spite of the fact that the latter also dealt, quite correctly, with statically indeterminate[2] problems, including the two-hinged arch (see Section 3 of this Chapter). In the following section, we shall revert to the solution of the problem as advanced by Coulomb who examined not only the equilibrium against slipping (having regard to friction and adhesion) but also the equilibrium of the individual voussoirs against tilting, and whose method was improved upon by Gauthey in his work on bridge building, and by Navier in his "Leçons".

In Germany, it was *Eytelwein* (1764–1848) who, in his "Handbuch der Statik fester Körper", dealt with the theory of vaults. Eytelwein was the founder and first Director of the Berlin "Bauakademie" which was the forerunner of the Berlin Technische Hochschule. He, too, went back to De la Hire, but took the friction at the joints into consideration.

But it was only when the elastic deformations of the arch were included in the calculation, that it was possible to create a theory of the fixed arch which satisfied the more stringent requirements. This approach to the problem which, in its origins, goes back to Culmann and Bresse, will be briefly reviewed in Chapter IX (page 220) in connection with the further development of the methods of structural analysis.

---

[1] "Si une voûte était composée de matériaux incompressibles . . . et qu' (elle) ne pût avoir absolument aucun tassement, il suffirait, pour qu'elle se soutînt, que les culées eussent une épaisseur convenable, et que la hauteur de la clef fût assez grande pour que la pierre ne s'écrasat" (Gauthey, page 194).

[2] For readers not familiar with technical problems, it may here be explained that a problem is called "statically indeterminate" if it cannot be solved according to the rules of the statics of rigid bodies alone. Thus, a beam resting on two supports is "statically determinate", as the reactions at the two supports can be determined by the moments' equation or by the laws of the lever. A beam resting on three supports, however, is "statically indeterminate", as it is necessary to take into account the *elasticity* of beam and supports in order to determine how the load is distributed on the three supports.

2. COULOMB

Among the French Génie officers of the eighteenth century, the one most important for the development of science was *Charles Auguste Coulomb* (1736–1806).

After completing his studies in Paris through which he acquired, in particular, a thorough mathematical knowledge, Coulomb was posted as a Génie officer to the French colony of Martinique where he was in charge of fortification works. On this occasion, he used the opportunity to investigate the solidity and statical behaviour of building elements, especially walls and vaults, and to deal with the problem mathematically. The results of these investigations which were, at the outset, merely destined for his personal use, were summarized in a treatise which he submitted to the Académie des Sciences and which appeared in the 1773 volume of the "Mémoires des Savants Etrangers", under the title "Essais sur une application des règles de maximis et minimis à quelques problèmes de statique relatifs à l'architecture".

In 1776, Coulomb returned to France where he continued to carry out thorough investigations into physical, technical and mechanical problems. In particular, he earned a reputation through his research on the elasticity of wires, and on electricity and magnetism. The torsion balance invented by him which is based on the torsional elasticity of metal wires, has gained importance in the technique of precision measurement. In 1784, he became a Member of the Académie des Sciences and, at the same time, Inspector of hydraulic works, retaining this office until the end of the monarchy in 1792. During the years of the revolution, he lived as a private individual until he was appointed, under Napoleon, a Member of the Institut National and Inspector-General of Public Education.

In the field of civil engineering proper, the great significance of the French scientist lies in the fact that, in dealing with problems concerned with statics and with the strength of materials, he used exact-scientific methods but was, at the same time, conscious of their application to building practice. As a physicist of distinction, he occupies a place in the history of science. As a theoretician, he is decidedly superior to his

Fig. 47. Dam at Lampy built in 1777–1781 to feed the Canal du Languedoc

Photo by courtesy of Secrétariat d'état à la production industrielle et aux communications, Canaux du Midi et Latéral à la Garonne

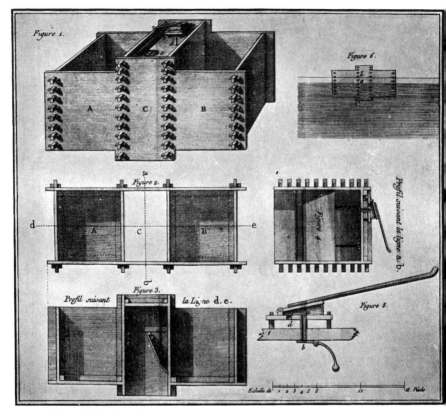

Fig. 48. Diving bell with floating chambers for under-water engineering works, according to Coulomb

predecessors and colleagues among the French engineers : Bélidor, Perronet, Gauthey, Chézy. But, in contrast to the "exclusive" scientists of the eighteenth century who, like the Bernoullis, Euler and others, were mainly interested in finding possibilities for the application of their new mathematical methods, Coulomb was, at the same time, an *engineer* who had to plan and supervise engineering works and was thus continuously confronted with the practical task of determining the dimensions of structures.

As far as the development of the methods of structural analysis is concerned, it is mainly the above-mentioned treatise from his Martinique period which must be regarded as a document of vital significance. In it, the classical problem of the bending of a beam has been finally solved, exhaustively and correctly, and even the problem of shearing stress is touched upon. In the same treatise, Coulomb also develops a thesis for the failure of compression-stressed masonry and brick structures, as well as the method for the calculation of earth pressure on retaining walls which is named after him and which, with a few perfections, is still being applied today. Lastly, the author also makes a contribution to the theory of vaulting which, even though it does not yet provide a final solution to the problem of the fixed arch, goes beyond the traditional method of approach as practised by de la Hire, Bélidor and others (cf. Fig. 49).

The publication of Coulomb's *Mémoire* ought to be regarded as a milestone in the history of structural analysis. Unfortunately, the rich contents of the treatise were compiled in such a concise form and concentrated into so little space that, as Saint-Venant points out, most of it escaped the notice of experts for forty years. This was all the more understandable as the author, in later years, occupied himself no longer with these problems but turned to other branches of physics.

The way in which the French engineer approached the solution of structural stress problems is apparent from his method for the determination of earth pressure, familiar to all civil engineers. Where complicated statical phenomena are concerned, the designer can have but an approximate picture of their effect. In the case of earth pressure, for instance, he

147

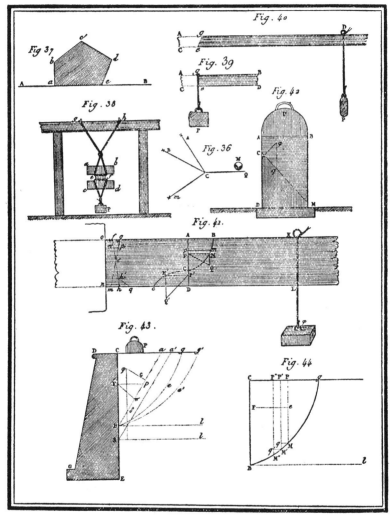

Fig. 49. Statical problems according to Coulomb.
From *Essai sur une application des règles de maximis et minimis.*

will assume that an earth prism of quasi-triangular cross-section attempts to detach itself from the remainder and to push the retaining wall outwards. But the individual factors which determine the magnitude of this effect, such as the size and shape of the earth prism, cannot be calculated precisely and accurately. Coulomb's method of approach is to let the unknown factor *vary* (e.g. the position or inclination of the plane along which the earth prism is assumed to slip), and to regard as the significant solution the *limit* value (maximum or minimum, as the case may be) which the wanted pressure may assume at the moment at which slipping (downwards or up-. wards) will occur. The maximum value will apply to active earth pressure, and the minimum value to so-called passive earth pressure, i.e. earth resistance.

The constants which Coulomb requires for his solutions of stress-analytical problems, are those indicating the *friction* and *cohesion* of the material. By the latter, he generally means both the shearing strength and the tensile strength, having found but little difference between them in the course of rupture tests with stone materials.

In a similar way, Coulomb examines the strength of a compression-stressed masonry pillar, varying the inclination of the joint along which rupture or slipping is assumed to occur. As is indeed confirmed by experiments with homogeneous, rigid material, the rupture will occur if the shearing strength is exceeded along an oblique rupture plane, inclined at an angle which is determined by the relation between inner friction and cohesion of the material concerned (witness the "double pyramid" apparent after cube crush tests).

Coulomb's approach to the theory of vaults is not much different. By way of introduction, he proves that, if friction and mortar adhesion are ignored, every vault must necessarily have the shape of an inverted catenary. He then turns to the examination of actual vaults which, owing to inner friction and cohesion, will be stable even if they deviate from the theoretical shape to some extent. For this purpose, he uses the friction and cohesion coefficients of the material to determine the *limit* values within which the horizontal thrust must be contained, if slipping or tilting at any potential disintegration joints is to be avoided. In other words, he determines the *greatest* thrust which will still allow slipping or tilting downwards at any joint (assumed to be normal to the intrados), and the *smallest* thrust which will already cause slipping or tilting upwards. The former value must be smaller than the latter. The greater their difference, the greater is the margin of the *actual* vault thrust lying between these extreme values, and the greater is the stability of the vault.

For the practical application of his method, the author suggests a number of simplifications. To begin with, the friction of the building materials normally used for vaults is so great that there is, in practice, no danger of sliding as between individual voussoirs. The first of the two balance conditions, relating to slipping, can therefore be neglected, and the examination can be confined to the case of tilting. In this case, however, the potential centres of rotation must not be assumed to lie on the edge of the cross-section. "It must be assumed that these points are sufficiently remote from the end of the joints so that the adhesion of the voussoirs will prevent these forces from breaking their corners ; this is determined by the methods which we have used in analyzing the strength of a pillar."[1]

In the concrete, individual case, the wanted limit values of the thrust can be determined with sufficient accuracy by applying the calculation to a few potential rupture joints in the vicinity of the quarter points. It is fortunate, in this connection, that the value of a function changes but little in the vicinity of the extreme points.

Rather incidentally, in connection with his observations on strength tests (tension, shearing and bending tests), Coulomb also deals, in his *Mémoire*, with the bending problem of a cantilever beam of rectangular cross-section. His treatment is universal in that he takes into consideration the shearing strength as well as the compressive and tensile strength, and admits, in principle, of any relationship between stress and strain. As a special case, applicable to the perfectly elastic body, he then obtains the well-known relation $M = \sigma \frac{bh^2}{6}$.

In conclusion, it can be stated that Coulomb, more than any of his predecessors, succeeded in dealing with statical

[1] "il faut supposer que ces points sont assez éloignés de l'extrémité des joints, pour que l'adhérence des voussoirs ne permette pas à ces forces d'en rompre des angles ; ce qui se détermine par les méthodes que nous avons employées en cherchant la force d'un pilier."

problems on a scientific basis but with full consideration of practical requirements. "I have endeavoured, to the best of my ability, to make the principles which I have used, sufficiently clear so that a technician with some training can understand and use them."[1] It is because of his endeavour to meet the mentality of the engineer and designer by laying emphasis on the perspicuity of the subject that Coulomb must be regarded as the first founder of *structural analysis*, or *building statics*.

Among Coulomb's later writings, there are mainly two which have a bearing on the history of civil engineering and which may here be briefly mentioned. In a *Mémoire* submitted to the Paris Academy in 1779, the author makes a concrete suggestion for the use of compressed air for engineering works under water. The wooden caisson proposed by him consists of three parts. The central part contains a working chamber with air pump and entrance opening. On either side is a floating chamber with sealed bottom so that the entire device can be floated to the site (Fig. 48, facing page 147).

In another, lengthier *Mémoire*, Coulomb proves a precursor of Taylor and other modern physiologists in that he examines, on a scientific basis, the effect and efficiency of human labour with different occupations and under different working conditions. In order to make rational use of the human body, that most versatile and suitable of all machines, one must aim at "augmenting the effect without augmenting the fatigue". If it were possible to express effect and fatigue by formulas, one must try to make the ratio of effect and fatigue (i.e. the "efficiency") a maximum. Having formulated the problem generally, the author proceeds to analyze observations regarding the carrying of heavy loads when climbing mountains ; statistics compiled by Vauban on transport of materials with wheel barrows, etc. It is interesting to note how, already more than 150 years ago, a keen observer recognizes the importance of breaks and recommends (quite independent of any social considerations) to exact not more than 7–8 effective working

---

[1] "J'ai tâché autant qu'il m'a été possible de rendre les principes dont je me suis servi assez clairs pour qu'un artiste un peu instruit pût les entendre et s'en servir."

hours per day in the case of heavy work. (With other, lighter occupations, however, even 10–12 working hours will not entail excessive fatigue.)

3. NAVIER

Louis Marie Henri Navier (1785–1836), whose name is familiar to engineers through the theory, named after him, of the consistently plane cross-sections, was born at Dijon, four years before the outbreak of the French Revolution. From 1802 to 1807, he studied at the Paris "Ecole Polytechnique" which had been founded in 1794 under the auspices of Monge and Carnot (the father of the well-known physicist), and subsequently at the "Ecole des ponts et chaussées". Then, only 22 years of age, he started his practical career as "Ingénieur des Ponts et Chaussées" of the Seine Département.

Orphaned at the age of 14, Navier's secondary education was entrusted to his uncle, Gauthey (cf. page 127), whose knowledge and intellect were to exert a decisive influence on the scientific development of the youth. Gauthey left to his nephew the manuscript of the "Traité de la construction des ponts", already mentioned, which was published by Navier in 1813.

Apart from practical engineering works (among his works are several bridges over the Seine, e.g. those at Choisy, Asnières, Argenteuil), Navier revealed an early inclination for scientific and pedagogic activities. Soon after the conclusion of his own studies, he joined the teaching staff of the "Ecole des ponts et chaussées", first as a deputy assistant, then in 1821 as an extraordinary professor, and finally, in 1830, as an ordinary professor for Applied Mechanics. In this capacity, he regarded it as his main task to apply the discoveries and methods of theoretical mechanics to practical tasks of construction, and to equip the engineering students with an appropriate scientific armour.

In the course of his teaching activities, Navier published numerous treatises, including, *inter alia* : "Sur la flexion des verges élastiques courbes" (1819) ; "Sur les lois d'équilibre et du mouvement des corps solides élastiques" (1821) ; "Sur les ponts suspendus" (1823).

Moreover, the first numbers of the *Annales des ponts et*

chaussées, that sterling French engineering journal founded in 1831, contain numerous contributions from Navier's pen, including reviews and minor reports as well as major essays on general civil engineering subjects, such as innovations in road construction in England,[1] or concessions and execution of public works, etc.

But of paramount importance for the development of civil engineering were the brilliant lectures which Navier held, at the "Ecole", on applied mechanics and structural analysis. They were repeatedly published in print, for the first time in 1826 under the title "Résumé des Leçons données à l'Ecole des Ponts et Chaussées, sur l'Application de la Mécanique à l'Etablissement des Constructions et des Machines". It is through these lectures that the theories of structural analysis and the strength of materials have become branches of engineering science in the modern sense. Their epoch-making value lies not only in the numerous new methods contained in the work ; of even greater importance was the fact that Navier, for the first time, integrated the isolated discoveries of his predecessors in the fields of applied mechanics and related subjects into a single, unified system of instruction, and that he taught his students how to apply the laws and methods already known to the practical tasks of structural engineering, i.e. to the determination of structural dimensions. In doing so, he became the actual creator of that branch of mechanics which we call *building statics*, or *structural analysis*.

A surprisingly great number of methods, still in daily use, of calculating the dimensions of structures go back to Navier, or have at least been re-formulated by him. They include, first of all, the classic theory of flexure, named after him ; a thorough examination of the problem of buckling, also for eccentric loads ; the systematic solution of a number of statically indeterminate[2] problems, such as the beam with one or two fixed ends, the continuous girder on three or more supports, the two-hinged arch, etc.; finally, some problems of the theory of flat slabs and plates. In an article published some years ago,[3]

[1] cf. Footnote 1 on page 164.
[2] See Footnote 2 on page 145.
[3] *Schweizerische Bauzeitung*, Vol. 116, page 201 (2nd November, 1940).

Professor F. Stüssi has pointed out the often amazing conformity between the results obtained with the aid of Navier's formulas and methods, and the results of recent tests or of modern, more exact methods of calculation, e.g. in the case of problems of buckling or flat slabs.

As far as the theory of flexure is concerned, the problem had already been solved by Coulomb fifty years earlier (see the preceding section) but Coulomb's success had not become known to a wider circle of engineers. Even Navier's first publications do not give the correct solution ; they still subscribe to Bernoulli's and Mariotte's assertion that the position of the neutral axis is indifferent. In 1819, Navier corrects his error partly so that he arrives at correct results for symmetric cross-sections. But it is only in 1824 that he formulates the modern terms (correct where stress and strain are proportional) for the deflection and ultimate strength of the beam subjected to bending.[1]

There is, however, one obvious difference between Navier's "Leçons" and a modern course of lectures on building statics, namely, the absence of graphic methods of determination and presentation, such as force polygon and funicular polygon, moment area, influence line and the like. On the other hand, the principal terms used by Navier for bending, buckling, etc., differ from modern terms merely in formal details. For instance, Navier does not yet use the present-day symbols for the moments of inertia and resistance, which express the purely geometrical characteristics of a beam. Instead, he introduces the conceptions of the "Moment of bending resistance" and "Moment of rupture resistance" in which the geometrical and physical characteristics of the beam are jointly expressed, and calculates them for a number of the most common cross-sections (double-tee, tee, channel, pipe, etc.).

As mentioned on page 76, the "Moment of bending resistance" had already been introduced by Jakob Bernoulli, a century earlier, as the reciprocal of the bending radius, and had been used by Euler, fifty years later, in his investigations of elastic lines. But whereas Bernoulli and Euler introduced the

---

[1] Saint-Venant, para. 9.

conception of "absolute elasticity" merely as a constant, characteristic for the beam concerned, Navier succeeded in calculating the actual numerical value from the cross-section of the beam and from the elasticity of the material.[1]

In dealing with the built-in beam, the continuous girder, the two-hinged arch and similar problems, Navier's method of approach is to formulate and integrate the differential equation of the elastic line in order to obtain an equation for the deflection whereupon he determines the integration constants from the support conditions. In the solution of other problems, however, such as earth pressure, stability of fixed vaults and the like, he follows Coulomb's methods. In any case, Navier attempts to provide the designer with the scientific aids which enable him to find a correct solution for the problems of the practice.

In the article already referred to, F. Stüssi sums up the importance and particularity of Navier in the following words : "The task which Navier set himself, is nothing less than the formulation of a proper method of structural analysis. . . . There is a difference in principle between him and his predecessors and contemporaries : Navier departs from experience and from building statics based on experiments, whereas others before him, and even after him, regarded building statics merely as a theory of equilibrium problems which need not be solved by experiment but solely on the basis of theoretical reflections. . . . In the most fruitful manner, Navier combined the mentality of the structural designer with a perfect mastery of the theoretical principles and aids. . . . His reflections are in keeping with an economic approach. The solution of the differential equation of the elastic line, for instance, is not only an attractive mathematical problem. . . . Its evaluation permits an appreciation of the elastic behaviour of a structure. . . . Navier's mentality as an engineer and designer is also apparent from his assessment of the validity limits of a theoretical result. . . . This clear perception of the

[1] The "modulus of elasticity" was first introduced in 1807 by the English physician, scientist and physicist Thomas Young (1773-1829), though not in the modern form of a stress (tons per square inch) but with the dimension of a length (inch) ; cf. Saint-Venant, paragraph 10 ; Todhunter, Chapter II.

essentials, this power of abstraction was a gift which, even after Navier, only few statical experts and designers possessed, such as perhaps Carl Culmann in his graphic statics and in his conception of the force composition in truss girders, and Otto Mohr in some of his treatises in the field of technical mechanics. All his successors, however, have been able to make creative contributions in certain fields only. For, Navier anticipated them all in the formulation of comprehensive methods of structural analysis.

"The fact that we are able, today, to construct safely and economically, is mainly due to the methods of structural analysis, that particular branch of mechanics which is based on the actual working conditions of a structure. These methods were created, within little more than a decade, by a single man, Navier."

Navier himself was conscious of the novelty of his procedure. In the preface to the first edition of his "Leçons", he states expressly that "statical examinations have hitherto been more useful for the progress of mathematics than for the perfection of the art of building".[1] Most designers determine the dimensions of building elements or machine elements according to experience, after the example of existing structures. "They seldom investigate the forces which these parts support, and the resistance which they offer."[2]

But it is not only as a creator of building statics that Navier must be regarded as a pioneer and precursor of the modern civil engineer : in his rich life, he became acquainted not only with the creative part of his profession but also with its tragic side. He had to carry a heavy load of that responsibility which is the inevitable lot of man in his struggle to master nature itself, and which has overshadowed the life of many a daring engineer. Owing to a combination of adverse circumstances : poor subsoil, difficult water drainage, jealousies and enmities within the Paris City Council, one of his principal works, the suspension bridge over the Seine in Paris (Fig. 50), was a

---

[1] ". . . ont été jusqu'à présent plus utiles aux progrès des mathématiques qu'au perfectionnement de l'art des constructions."

[2] "Ils se rendent compte rarement des efforts que ces parties supportent, et des résistances qu'elles opposent."

failure. The structure, already almost completed, had to be dismantled in the end, and the matter cast a tragic shadow over valuable years of Navier's life, in spite of the fact that, according to the judgment of the most earnest contemporary experts,

Fig. 50. One of Navier's projects for the suspension bridge over the Seine, Paris. Anchoring of the chain.
From Navier, *Mémoire sur les ponts suspendus*.
Reproduced from *Schweizerische Bauzeitung*, Vol. 116.

Navier was free from blame. *Prony* (1755–1839), co-founder and leader of the "Ecole Polytechnique" for many years, calls the case "one of the more or less serious accidents which the engineer often encounters in his great works." Navier, alluding to this matter, writes : "To undertake a great work, and especially a work of a novel type, means carrying out an experiment. It means taking up a struggle with the forces of nature without the assurance of emerging as the victor after the first attack."[1]

4. THE DEVELOPMENT OF THEORETICAL MECHANICS

Owing to Navier, *Structural Statics* was thus established as a special branch of science. Thanks to the ceaseless research work of untold, able engineers with theoretical backgrounds, the methods of structural analysis have since been steadily perfected and extended to new tasks, and to other building materials. All the time, however, the character of an *applied* science, subservient to practical purposes, has been preserved.

This was not the case with *theoretical mechanics*. As a branch of physics, this was mainly cultivated and advanced by mathematicians and by a few engineers of exceptional mathematical ability. True, these continued to provide most valuable inspiration to structural designers. But, generally speaking, the gulf between pure science and applied science widened more and more. For the purposes of the present survey, we can but briefly deal with a few elementary problems and research results within the vast domain of theoretical mechanics and the mathematical theory of elasticity, which have a particular bearing on the science of structural analysis and of the strength of materials.

From Galilei's and Mariotte's first experiments to the final solution submitted by Coulomb and his successors, the elementary theory of flexure was based on the assumption of a moment-resisting beam consisting of individual, parallel *fibres*, an assumption which had naturally emerged from the observation of the wooden beam. In other words, the examination was largely confined to the one-dimensional state of stress or, if shearing forces were included, to the two-dimensional state of stress which is generally sufficient for the practical task of determining structural dimensions.

Already since the eighteenth century, however, a number of physicists and mathematicians have attempted to analyze

---

[1] "Entreprendre un grand ouvrage et surtout un ouvrage d'un genre nouveau, c'est faire un essais ; c'est engager avec les forces naturelles une lutte dont on n'est point assuré de sortir vainqueur dès la première attaque." (Quoted by F. Stüssi, *Schweizerische Bauzeitung*, Vol. 116, page 204, 2nd November, 1940).

the state of stress in a solid body on a *three-dimensional basis*. Even Newton had expressed the opinion that the smallest particles of matter attract each other in the same way as celestial bodies are attracted by gravitation.[1] The Jesuit padre and mathematician *Boscowich* (1711–1787), a native of Ragusa in Dalmatia, who was later mostly resident in Italy, and who has already been mentioned in connection with the statical examination of S. Peter's dome, was one of Newton's first followers and supporters in Italy, and ascribed the cohesion and elasticity of solid bodies to molecular forces. He visualized the smallest particles of matter as dimensionless points which are, however, endowed with repelling and attracting forces, and which are thus able to build up dimensional, impenetrable, coherent and elastic matter.[2] According to him, nature does not, therefore, know any absolutely rigid body.

This theory of the "molecular forces", as the inner stresses were then called, was amplified and mathematically treated by numerous scientists, including *Poisson* (1781–1840) and mainly, *Navier*. "New in Navier's thesis was that he, on the assumption that the inner (molecular) forces are in equilibrium prior to the deformation, remained content to consider their increase or decrease only. Generalizing the principle propounded by Hooke and Mariotte, he assumed that these increases or decreases are proportional to the change of distance between the molecules." By means of an elegant application of Lagrange's methods of analytical mechanics, Navier succeeded in formulating stress equations for isotropic bodies, although he does not yet use the term "stress" himself.[3]

The actual foundations of the mathematical theory of elasticity, and of the general theory of the strength of materials, are however mainly due to *Cauchy* (1789–1857) who, like Navier, had received his scientific training at the "Ecole Polytechnique" and at the "Ecole des ponts et chaussées". As a young engineer, he was, for a short time, engaged in harbour works at Cherbourg. But after no more than four years, he

---

[1] Todhunter, Chapter I ; also Saint-Venant, para. 22.

[2] cf. Rosenberg, "Geschichte der Physik", Vol. II ; also Saint-Venant, paragraph 22.

[3] Saint-Venant, paragraph 23.

abandoned the practical work, uncongenial for him, and returned to Paris so as to be able to devote himself entirely to science,
and particularly to pure and applied mathematics. After the
political upheaval of 1830, he lived in exile for eight years, first
at Fribourg in Switzerland, and later at Turin where Carlo
Alberto had offered him a professorship of physics.

As a scientist, Cauchy was extremely fertile. His work
comprises nearly 800 treatises which deal with practically all
branches of pure and applied mathematics. In particular,
Mechanics and the theory of the strength of materials were
treated on a strictly mathematical basis, and not so much with
a view to the immediate requirements of practical application,
which Navier never lost sight of.

Like Poisson and Navier, Cauchy departs from the assumption of attracting or repelling forces in the individual molecules,
and arrives at the notion of "stress" through a summation of
the individual forces. In order to analyze a state of stress prevailing inside a body, Cauchy visualizes a small elementary
particle, e.g., a tetrahedron or a right-angled or oblique-angled
parallelepiped, cut out from the body, and proceeds to examine
the conditions of equilibrium so as to derive, from them, the
relations between the stresses prevailing at the different faces.
The analysis of "Cauchy's Tetrahedron" leads to the well-
known equations which permit the reduction of the inner
stresses at any cross-section of the body to the stresses prevailing in three normal planes intersecting at the same point.
The equilibrium conditions of the parallelepiped yield the
theorem that, in the case of two inclined intersecting planes,
the component of the first stress normal to the second plane, is
equal to the component of the second stress, normal to the first
plane. In the special case of the rectangular parallelepiped,
this theorem leads on to the well-known *theorem of the shearing
stresses* according to which the shearing stresses in two normal
intersecting planes must always be equal. From the equilibrium
conditions of these elementary particles, Cauchy also derives
the three *basic elastic equations* :

$$\frac{\delta\sigma_x}{\delta x} + \frac{\delta\tau_{yx}}{\delta y} + \frac{\delta\tau_{zx}}{\delta z} + X = O$$

and similarly, *mutatis mutandis*, for the other two co-ordinates.

Through relating the normal and tangential stresses inside an elastic body with the strains and displacements caused by them, Cauchy derives further basic equations.

Cauchy is also responsible for the method whereby the state of stress or strain is presented as a *stress ellipsoid* or *strain ellipsoid*, the axes of which coincide with the direction of the primary stresses or strains.

Cauchy's essays on the mathematical theory of elasticity are mainly to be found in the several volumes of his work, "Exercices de mathématiques" (1827–1829). This book contains, in addition to the basic theorems, equations and notions, since reproduced in all textbooks on the theory of the strength of materials, a number of complicated investigations of a theoretical nature which are outside the normal sphere of interest of the practical engineer.[1]

In addition to Cauchy, certain contemporaries and compatriots of his played an important part in the further advance of theoretical mechanics. They were *Lamé* (1795–1870), *Clapeyron* (1799–1864) and *Poisson* who has already been mentioned. The latter, in particular, was the first to note, in 1827, that a prismatic rod, subjected to an axial tensile stress, shows a contraction in the transverse direction. Lamé and Clapeyron are also responsible for a method, different from Cauchy's, of representing the state of stress by an ellipsoid, in which the vectorial radii represent the magnitude and direction of the stresses. Furthermore, Clapeyron's name is familiar to every civil engineer from the "theorem of three moments", applicable to the continuous beam. This theorem, often named after him, was formulated on the occasion of the construction of the great railway bridges over the rivers Seine, Garonne, Lot and Tarn.

The mathematical theory of elasticity, created by the French school, is based on the assumption of proportionality between the stresses and the elastic deformations caused by them (Hooke's Law). The bodies are regarded as isotropic, and it is assumed that the deformations remain very small so that it is sufficient to consider the first term of the series. Through

[1] cf. Todhunter.

applying the theory to simple cases which permit the integration of the differential equations concerned, the mathematicians named in this section were already able to solve, at that time (about 1820-1850), a number of special problems which were to acquire practical importance in the field of civil engineering only at a much later time, such as the behaviour of flat slabs and membranes, torsion and flexure of solid and hollow cylinders (pipes), hollow spheres, etc. *Saint–Venant* (1797–1886) who made valuable contributions of his own, e.g. to the problem of torsion, has presented a detailed history of this extremely fertile early period of theoretical mechanics in the historical introduction to the third edition of Navier's "Leçons", edited by him.

CHAPTER VII

THE INDUSTRIALIZATION
OF EUROPEAN CULTURE

I. THE INDUSTRIAL REVOLUTION IN ENGLAND :
COAL, STEAM ENGINE AND RAILWAYS

Whilst France occupied a leading position among the
nations of Europe as regards the exact sciences and their
applications to technical problems, a process of world-historical
importance took place on English soil during the eighteenth
century, a process known as the "Industrial Revolution".
During the second half of the eighteenth century, "the civilization of the riding-horse and the pack-horse gave way to that
of the coach, the waggon and the barge",[1] and the centuries-old
soft roads, dusty in dry weather and mud-bound during rain,
were replaced by hard roads and canals.

The part played by Brindley in the construction of British
canals has already been referred to (cf. page 134). The improvement of the roads is inseparably linked with the name of
Macadam.

From his twenty-sixth year on, the Scottish engineer, *John
Loudon Macadam* (1756–1836), devoted his life to problems of
road construction and road maintenance. Having carried out
his first studies and experiments at his own expense, he was
soon generally recognized as an authority on road engineering.
As Road Inspector of the "Bristol Turnpike Trust", he undertook the reconstruction and maintenance of the roads of his
district in accordance with his own principles, with the result
that the maintenance costs were soon considerably reduced.

[1] cf. Introduction to Trevelyan's "British History in the Nineteenth
Century".

Since 1817, the much frequented approach roads to the Thames bridges in London were renewed in accordance with his system.

Macadam recognized that the road surface must, first of all, be protected against the destructive soaking effect of water. This he achieved by the strict observation of the following principles : The foundation of the roadway should rise at least 3–4 in. above the water level in the soil, or in the ditches. The roadway proper must be formed of a compact 6–8 in. layer (later two layers), consisting of 1½–2 in. diameter broken stone which must be completely free from earth admixtures. The convexity of the cross-section must be approximately 3 in. at the centre. During the period of consolidation, the road surface must remain smooth and even. The ruts, which are necessarily caused by the first carriages passing over the freshly put on pavement, must therefore be raked in at once. Later on, too, any damage to the road surface must be continuously repaired.[1]

The new roads with their hard and smooth surfaces permitted a considerably speedier, heavier and more economic traffic on wheels. The consequence of this development for the general economy of the country can hardly be overestimated. The expression "Macadam" became not only a technical term but "a symbol of all progress and was metaphorically used in common parlance for any aspects of the new age where improved and uniform scientific methods were in demand".[2]

The intensification of goods traffic, made possible by the new roads and inland waterways, was the first prerequisite for the replacement of the long-accustomed home-industry by factory production. We are here not concerned with the tremendous growth of the mechanized textile industry, particularly the cotton industry,[3] which multiplied its production a hundred times between 1760 and 1830 ; nor are we here concerned with the profound social consequences of this precipitate industrialization. Another aspect of the industrial

---

[1] cf. Navier, "Considérations sur les travaux d'entretien des routes en Angleterre", in *Annales des ponts et chaussées*, Paris, 1831, Vol. II, page 132.

[2] cf. Trevelyan, "British History in the Nineteenth Century", page 166.

[3] The first spinning machine was built in 1770 ; the invention of the mechanical loom followed in 1786.

revolution, however, was of the greatest indirect importance to building construction and thus to the civil engineer, namely, the technical development of the mining industry which led to a tremendous increase in the extraction of pit coal.

Since the Middle Ages, the output and consumption of pit coal had been greater in England than in any other country of Europe. Already during the thirteenth century, domestic coal consumption in London is said to have been so great that restrictive bye-laws became necessary to check the increasing smoke nuisance. During the seventeenth century, English coal was already shipped to the Continent in considerable quantities.

The actual "coal age", however, set in during the second half of the eighteenth century when it became possible to use steam power for the drainage of collieries which permitted the working of deeper galleries under conditions of greater safety. Most of the achievements of that time belong to the history of mechanical engineering : the feat of *Papin* (1647–1712) who invented the atmospheric piston engine and built a small-scale model of it ; the achievements of *Savery* (approximately 1650–1715) and *Newcomen* (1663–1729) who were the first to apply these engines to the practical purpose of colliery draining (around 1712) ; the ingenuity of *Smeaton* (cf. page 134) who made considerable improvements on the atmospheric engine, through a better choice of proportions and improved workmanship of the component parts, and who constructed some such machines of very large dimensions.[1] We must also be content with a mere mention of *James Watt* (1736–1819) whose inventive genius developed the "fire engine" of his predecessors into the steam engine proper, embodying such features as a condenser independent of the cylinder, the transformation of the reciprocating movement into a rotary movement, a centrifugal governor and other vital improvements.

The utilization of steam power led to an enormous increase in the extraction of cheap coal which reacted on the progress of civil engineering in two ways : through the progress of metallurgy at large, which will be the separate subject of the following

---

[1] One of his engines, installed in a Cornish colliery, had a cylinder of 6 ft. diameter and 10½ ft. height (Matschoss). Some of the pioneer engines mentioned here are exhibited at the Science Museum, London.

section, and through the advent and growth of *railways*. The capacity and efficiency of the new means of transport dwarfed all earlier ones and dominated the entire civilization of the nineteenth century. The construction and maintenance of railways represented one of the foremost, if not the most important, task of civil engineering during that century and gave rise, in addition, to the construction of numerous important buildings and bridges.

The railways developed from the "tramroads" or "tramways" which were used in English mining districts, since the seventeenth century, for the transport of coal from the pit-heads to the ports.[1] On these tramroads, the same tractive

Fig. 51. Cast iron rail, shaped as a beam of uniform resistance.

force was able to haul two-and-a-half times as much coal as on ordinary roads. A report from 1676, quoted in the "Encyclopedia Britannica", gives the following description : ". . . rails of timber laid from the colliery to the river, exactly straight and parallel, and bulky carts were made with four rollers fitting the rails, whereby the carriage was so easy that one horse would draw down four or five chaldrons of coal."

The first rails consisted of wooden beams, mostly of oakwood, which were, however, soon lined with iron plates or strips for protection. The next step was to replace the plates by a kind of angle iron so as to obtain an automatic guidance of the wheels.

Further progress was achieved when, towards the end of the eighteenth century, the guiding function was transferred from the rail to the wheel, by using *flanged wheels*. At the same time, the composite iron and timber rails began to be replaced by fishbelly-fashioned cast iron rails, shaped as beams of uniform resistance (Fig. 51). These were fixed on stone blocks by

[1] Similar tramways, with wooden "rails", have apparently been used in Continental coal mines (Harz Mountains, Ore Mountains, Tyrol) since the sixteenth century.

means of nails and wooden dowels, or on transverse sleepers by means of cast iron "chairs". The brittleness of the material, however, gave rise to accidents owing to broken rails, which only ceased when the cast iron rails were replaced by those of malleable iron during the first decades of the nineteenth century.

At the turn of the century, steam engines were in common use for pit drainage purposes in the British mining districts. It was a rather obvious idea to extend the use of steam power to other purposes, especially as a substitute for the tractive power of horses for the transport of coal. *Richard Trevithick* (1771–1833), who had already earned a reputation as a designer of high pressure steam engines which were lighter and more mobile than Watt's low pressure engines, and who had already used one of these engines to drive a road vehicle, built the first steam locomotive in 1804. This locomotive ran on a Welsh colliery tramroad and was able to haul a load of 13 tons at a speed of 5 miles per hour. Soon, further colliery tramroads adopted steam operation. On the tramway of Killingworth Colliery, a tender locomotive of 10 tons weight hauled a load of 40 tons at a speed of $5\frac{1}{2}$ miles per hour. On the Hetton tramroad, near Sunderland, opened in 1822, trains of 60 tons covered the 7 miles line from the pithead to the River Wear at a speed of $4\frac{1}{2}$ miles per hour. On the level sections, the trains were hauled by the locomotives. On sections with heavy gradients, stationary steam engines were used. It was for the Killingworth tramroad that *George Stephenson* (1781–1848) had built his first steam locomotive, "Blücher". He also supplied the first five locomotives for the Hetton Railway.

The obvious advantages of the steam-operated coal tramroads soon led to the idea of using the new tractive power also for the transport of other goods, and of passengers. The first railway destined for public transport was the Stockton and Darlington line of 25 miles length, built in 1823–1825. The undertaking was due to the initiative of *Edward Pease*, a colliery owner and contractor who entrusted George Stephenson, already well known by that time, with the design and construction of the line. The company founded by Pease was empowered by an Act of Parliament to operate the railway for public transport, and to haul the wagons by men, horses or "otherwise". The first railway ticket was sold on September 27th, 1825, and this date is regarded as the actual birthday of the railways which were soon to dwarf the traffic importance of Brindley's canals and Macadam's highways.

Although, on the Darlington-Stockton line, the transport of coal continued to be the main purpose and the most important source of revenue (representing seven-eighths of the total revenue), this first public railway roused the keenest interest. "New railway projects sprang into existence everywhere. The most important scheme among them was the proposal to connect the great port of Liverpool with the great commercial and industrial centre of Manchester. Merchants and contractors in both towns had recognized the need for efficient means of transport from their own experience".[1] True, the two towns were already connected by canals, but these were no longer capable of handling the traffic which had grown enormously owing to the rapid development of the textile industry. Cotton, which had arrived at Liverpool from America, could not be conveyed to Manchester quickly enough. Sometimes, the factories were at a standstill for days on end, because they were short of cotton. The most prominent among the factory owners therefore joined forces to promote the construction of a railway before Parliament. Many plans were afloat. But a man was needed who would be able to assume the responsibility for this great work. The Stockton-Darlington railway was studied, and George Stephenson was believed to be the only man capable of performing this task.

"At the outset, people did not believe that iron carriages would be able to move on iron rails, not even on level track. The friction would be much too small. It was then found that such motion was possible after all. But gradients would, of course, be impracticable. Even Stephenson tried to do without gradients, if possible, and to design the route correspondingly, even at the expense of great detours, costly tunnels and viaducts". With the exception of a short section where additional tractive power was provided to overcome a ramp and counter-ramp of approximately 1 in 96, the gradient nowhere exceeded

---

[1] This and other quotations in this section are from Matschoss.

168

1 in 880 over the entire line from the upper mouth of Liverpool Tunnel to Manchester. In order to achieve this end, extensive earth movements were necessary, at the cost of nearly £200,000. The great cutting at Kenyon alone called for the excavation of 800,000 cubic yards. Further difficult engineering works were required for the crossing of the notorious Chat Moss, and for the construction of Edge Hill Tunnel, Liverpool. Here, exceptional difficulties had to be overcome in dealing with certain sections exposed to the earth pressure of soft blue clay or wet sand so that the responsible engineer was often constrained to harangue his discouraged workmen in person, and to spur them on to continue. The engineering feats of the line comprised no fewer than 63 bridges of all types, including the great Sankey Viaduct, built mainly of brick and partly of cut stone, which crossed the valley in a flight of nine arches of 50 ft. span, about 70 ft. above the water level of the Sankey Canal. The cost of this viaduct alone amounted to over £45,000.[1]

"It was on a level 2 mile section of the Liverpool-Manchester line, at Rainhill, that on October 28, 1829, the famous trial took place which was to determine the future system of operation. Stephenson's "Rocket", reaching an average speed of 13 miles and a maximum speed of $21\frac{1}{2}$ miles per hour, exceeded all expectations. The "Rocket's" output was about 12 h.p., which was soon increased to 20 h.p. by means of certain technical improvements. Fuel consumption was 17–20 lbs. of coke per h.p.-hour. That was a great step forward compared with the earlier locomotives which had an output of 10 h.p. at the most, and consumed 28-31 lbs. of coal per h.p. hour. The Locomotive trials at Rainhill and the opening of the Liverpool-Manchester line on September 15, 1830, finally determined the question of locomotives *versus* stationary steam engines in favour of the former.

"The opening of the railway by the Duke of Wellington, famous General of Waterloo, and Prime Minister of the day, took place with great ceremony. All expectations were far exceeded. Traffic between Liverpool and Manchester increased

[1] cf. H. Booth, "Chemin de fer de Liverpool à Manchester" in *Annales des ponts et chaussées*, Paris, 1831, Vol. I, page 1.

enormously. Contrary to prior apprehensions, the value of the lands adjacent to the line did not decrease but increased considerably. The line was first operated with eight of Stephenson's locomotives. The management had hoped to be able to convey 400–600 passengers a day ; the actual figure was 1,200. By 1835, 500,000 passengers had already been conveyed. In 1830, the 30 mile journey from Liverpool to Manchester took just over an hour."

For his railways and locomotives, Stephenson chose the traditional gauge of the old colliery tramways which thus became the "standard gauge" throughout the world. During the first stage of railway development, however, the predominance of this gauge was far from being unchallenged. Thus, when *Isambard Kingdom Brunel* (1806–1859) became engineer of the Great Western Railway in 1833, he introduced the broad gauge of 7 ft. Eventually, up to seventy different gauges were in use. But, as and when the individual lines and spurs grew into a single network, it became imperative to standardize the gauge, and Stephenson's gauge, which was, at that time, already used by most of the main lines, gradually ousted all the others.

The British engineers chiefly responsible for the origin and growth of railways were, in many cases, civil engineers and mechanical engineers in one. In the same way as the civil engineers of the Italian Renaissance had been architects and, frequently, artists at the same time, and in the same way as many of the French "Ingénieurs des ponts et chaussées" had earned a reputation as physicists and mathematicians, it was typical for the technicians of the British "Industrial Revolution" and of the early "Railway Age" to be active both as civil engineers and mechanical engineers. Indeed, some of the most famous among them were originally mechanical engineers. Brindley, who had worked his way up from mill designer and mechanic to become the great canal builder, has already been mentioned (page 134). His collaborator, Smeaton, has not only improved the steam engine ; he also built the Eddystone Lighthouse, "a masterpiece of contemporary engineering, famous far beyond the British Isles". Trevithick, the designer of the first steam locomotive, also undertook the construction of a tunnel under the Thames which could not be completed, however, owing to the enormous technical difficulties, insuperable at the time. In 1832, one year before his death, he planned the construction of a cast iron tower of 1,000 ft. height.[1]

George Stephenson, who started work at a colliery as a boy of eight years, and became an auxiliary fireman at the age of 14, and engineman some years later, not only designed steam engines and locomotives and founded, with two partners, the first locomotive works of the world. He also acted, as we have seen, as a civil engineer responsible for the design and construction of the Stockton-Darlington and Manchester-Liverpool railways, with their great engineering works, including tunnels, bridges and viaducts. His son, Robert Stephenson, collaborated with his father in the design and manufacture of locomotives at the Newcastle Works. At the same time, however, he created epoch-making civil engineering works in the field of iron bridge construction. His "Britannia Bridge" will be referred to in the following section.

John Rennie the Elder (1761–1821) was concerned, around the turn of the century, with the construction of canals, the drainage of the Fens and the building of bridges, including the famous old Waterloo Bridge in London[2] built 1811–1817 in classicist style and demolished 1935. But he had started his career as a mechanic, especially as a mill designer, and he was the first mechanical engineer to use gear wheels of precision-cut iron instead of the conventional wooden ones.

These few examples—their number may be easily increased[3] — may suffice to characterize the British engineers of that epoch. Mostly emanating from the practice, they were less concerned with theoretical, scientific problems than their French colleagues of that time. But they were consummate

---

[1] cf. Matschoss, page 154.

[2] The existing "London Bridge", too, is based on the elder Rennie's design. But it was only constructed in 1825–1831 under the supervision of his sons, John and George, after their father's death. The bridge was enlarged in 1902–1904, but the faces with the classical stone cut were preserved.

[3] e.g. Marc Isambard Brunel, cf. footnote on page 194. Even James Watt, the famous inventor of the steam engine, was occasionally active as a civil engineer, e.g., when he supervised the construction of the Monkland Canal, about 1770 (cf. Beck, Vol. III, page 516).

masters of the entire sphere of technique, and engineers of daring and genius who made extremely important contributions to the art of engineering. However, the application of mathematics and statics to technical problems, as well as the creation of the modern methods of structural analysis and of the foundations of hydraulics, were largely the work of French engineers with scientific training, and it took a remarkably long time till the results obtained by them became common knowledge among British technicians as well. The unfortunate controversy between the followers of Newton and Leibniz regarding the priority of the discovery of the Infinitesimal Calculus had become inflated into a matter of national prestige, and had encumbered the relations between scientists on either side of the Channel for a century, preventing closer collaboration. As late as 1805, Robison wrote an article on the strength of materials for the "Encyclopedia Britannica" in which he dealt with the problem of flexure in accordance with Galilei's theories, completely ignoring the achievements of Mariotte, Parent and Coulomb. *Moseley* (1802–1872), Professor at Oxford University, acquainted his countrymen, whose scientific standard had somewhat lagged behind that of the French, with the works of Navier and Poncelet on building mechanics.[1]

The achievements of French scientists like Prony, Bellanger and others in the field of hydraulic engineering were first made known to the majority of British technicians through the agency of an American engineer, *Charles S. Storrow* (1809–1904) who had, in 1830–1832, amplified his studies at the "Ecole Polytechnique" and "Ecole des ponts et chaussées" in Paris and, on his return to America, published "A Treatise on Waterworks for Conveying and Distributing Supplies of Water".[2]

Even in the sphere of mechanical engineering, in which British engineers were unquestionably ahead, the difference in the mentality of British and French engineers is apparent. The British had been using steam engines for decades, without bothering to make theoretical investigations about them.

---

[1] Mehrtens, "Vorlesungen über Ingenieurwissenschaften", second edition, Vol. III, First Part, Leipzig, 1912, page 48.

[2] Boston, 1835 ; cf. *Engineering News Record* of 24th September, 1931, page 476.

Hardly had the new invention reached France, when Carnot enunciated the Principle of Reversibility which led to the discovery of the Second Law of Thermodynamics.

2. A NEW BUILDING MATERIAL : IRON AND STEEL

As already mentioned, the development of the technology of iron and steel production is closely related to the "Industrial Revolution" in England, and in particular to the increased coal output. At the time when French engineers and scientists created the theoretical foundations of civil engineering, and German and Swiss builders brought the art of timber bridge construction to a high degree of perfection,[1] Britain was the first country to produce the new building material in large quantities and greatly to improve its quality.

The high price of the material called, as it still does, for great economy, causing the designers of great iron structures to determine the dimensions of building elements as accurately and as economically as possible, on the basis of a structural analysis. On the other hand, elasticity and strength of the metal were much more uniform than those of stone material and, therefore, facilitated the application of the theories of structures and strength of materials, permitting a much higher ratio of safety stress and breaking stress. As has been pointed out with some justification, iron is a *constructional* material rather than a *building* material. But, for that very reason, it proved to be one of the mightiest sources of inspiration and progress in the field of civil engineering.

During the eighteenth century, "the English iron and steel producing industry succeeded in replacing firewood, which had become scarce, by pit coal, which was available in abundance.[2] In connection with the great achievements in this

---

[1] Bridges constructed by the master builder and carpenter, *Johann Ulrich Grubenmann* of Teufen (1709–1783), reached spans of over 300 ft. Cf. page 125 for works in the German language, dealing with the theory and practice of timber bridge construction.

[2] The smelting of iron in a blast furnace, using coal or coke, was first practised by D. Dudley (1599–1684), but the process remained comparatively unknown until it was re-invented a hundred years later (around 1735), by Abraham Darby II (1711–1763).

field, we remember the foundry masters, *Abraham Darby* (father and son) and that famous foundry engineer, *John Wilkinson* . . . But of particular significance was the invention of *Henry Cort* (1740–1800) who produced wrought iron in coal-fired flame furnaces through the so-called puddling process (1784), at a time when Watt's first steam engines began to transform the various branches of industry. If it is recalled that, prior to Cort's invention, British malleable iron was of such inferior quality that its use was not allowed in the Navy, and that Swedish and Russian iron had to be used instead, the significance of this invention will be apparent. In 1786, Lord Sheffield went as far as to express the opinion that the inventions which England owed to James Watt and Henry Cort, would more than offset the loss of America ! For, owing to the achievements of her engineers, England would be superior to all countries in the production and metallurgical treatment of iron."[1]

It was thus in England that iron was first used as a *building material* on a major scale. In 1777–1779, *Abraham Darby III* (1750–1791) constructed the famous 100½ ft. cast iron arch bridge over the Severn at Coalbrookdale which is still in existence today, without having undergone any structural changes (Fig. 52), though it is now being used as a footbridge only. During the following decades, a number of other cast-iron bridges were erected in England, including some of considerable dimensions, like the 180 ft. span of the bridge at Staines. On the Continent, the first structure of this kind was a footbridge of 43 ft. span which a wealthy estate owner in Lower Silesia erected across the Striegauer Wasser near Laasan, in 1797. In France, similar structures were built soon after the turn of the century, thus the Pont des Arts (1803) and the Pont d'Austerlitz (1804–1806) over the Seine in Paris.[2] The latter bridge was, in 1854, replaced by a solid structure.

All these bridges were arch-shaped ; their main girders consisted of individual cast iron pieces which formed either

[1] Matschoss, page 103 *et seq.*
[2] As early as 1755, the construction of an iron bridge had been planned at Lyon. In fact, one of the three 82 ft. arches had already been assembled at the workshop when it was decided, for economic reasons, to forego the use of iron, and to erect a wooden bridge instead (Gauthey, "Traité de la construction des ponts", 1843 edition, Vol. II, page 101).

bars or trusses like those of a timber structure, or a kind of wedge-shaped "voussoirs", acting as a vault. In the latter case, the individual pieces consisted of frame-like, openwork castings so that, in either case, a "spider-web" appearance was obtained. The difficulty consisted in jointing the individual castings with each other. This was done in different ways, e.g. by means of wrought-iron arcs (bridge over the Wear at Sunderland, built 1793–1796) or by means of groove and tongue (Thames bridge at Staines, built 1802). It was mainly due to these difficulties that reverses were experienced. In one case, the arch collapsed when the centreing was removed. The bridge at Staines, where the wedge-shaped castings were held together by means of groove and tongue, failed less than twenty years after its erection, after many unsuccessful repairs.

Whilst the cast iron arch bridges were still "manufactured" more or less in accordance with traditional craftmanship, it was easier to analyze the action of forces, and thus to apply scientific methods of calculation, in the case of *suspension bridges*. In China, chain-suspended bridges had been in use since time immemorial. In Europe, the first design for a suspension bridge can be found in a treatise entitled "Machinæ novæ" and published, in 1617, by one *Faustus Verantius*, a native of Dalmatia who later came to live in Venice. In the seventeenth and eighteenth centuries, emergency bridges suspended on ropes were sometimes used during military campaigns. The first modern suspension bridge was erected by *J. Finley* in North America in 1796. By 1810, the number of suspension bridges in the United States, built to the catenary principle, was already considerable. Of these, the 240 ft. bridge over the Merrimac River (Massachusetts), built in 1809, is still in existence today, the chains having been replaced by parallel-stranded wires in 1909.[1]

In Britain, too, important structures of this type were built at the beginning of the nineteenth century. A pioneer of suspension bridge building in Britain was *Samuel Brown*, a naval captain and engineer who replaced the forged chain links by flat iron bars, placed on edge and joined by bolts.

[1] cf. Schaechterle and Leonhardt, "Hängebrücken", in *Die Bautechnik*, 1940, page 377 *et seq.*

Brown was responsible for quite a number of chain suspension bridges, including the Union Bridge across the Tweed at Berwick, built in 1819–1820, with a main span of 442 ft. Famous are the Menai Suspension Bridge at Bangor, with a catenary span of 579 ft., and the Conway Castle Bridge (Fig. 53), with 416 ft. span, both built, between 1819 and 1826, by *Thomas Telford* (1757–1834), one of the most eminent British civil engineers of his time. Navier's construction of a suspension bridge over the Seine in Paris has already been referred to (page 156).

All these bridges were *chain* suspension bridges of wrought-iron. The first *cable* suspension bridge was erected in 1816, in North America. The first major cable suspension bridge in Europe was built at Geneva in 1822–1823. The designers of this bridge, which had two spans of 131 ft., were the Swiss engineer *Henri Dufour* (1787–1875) and the French engineer, *Marc Séguin*. The latter was, during the following decades, also responsible, together with his brothers, for the construction of a great number of important structures of this type in France.[1] During the years 1832–1834, the French engineer *J. Chaley* built the famous "Grand Pont" across the valley of the Sarine at Fribourg, Switzerland. The bridge, demolished in 1923, had a span of 896 ft. and was described by Saint-Venant as "the boldest structure ever built" (Fig. 54). Originally, the wooden platform which was nearly 22 ft. wide, including the two lateral footwalks, was carried by four cables, each consisting of 1056 wires of $\frac{1}{8}$in. diameter. In 1881, the bridge was strengthened by two additional carrying cables.

Most of the suspension bridges built during the first decades of the nineteenth century had no bracing, and not a few of them have failed after a comparatively brief existence, owing to the resonance vibrations caused by wind forces. The Union Bridge at Berwick-upon-Tweed, for instance, was destroyed during a gale only six months after its completion.

If the calculation of the additional stresses caused by aerodynamic forces represents a difficult problem which has still

---

[1] cf. Vicat, "Ponts suspendus en fil de fer sur le Rhône", *Annales des ponts et chaussées*, 1831, Vol. 1, page 94. Also, Séguin, "Pont suspendu en fil de fer à Bry sur Marne", *Annales des ponts et chaussées*, 1832, Vol. 1, page 210.

Fig. 52. Bridge over the Severn at Coalbrookdale, built in 1776–1779 by Abraham Darby

Photo by courtesy of the Director of the Science Museum, South Kensington, London

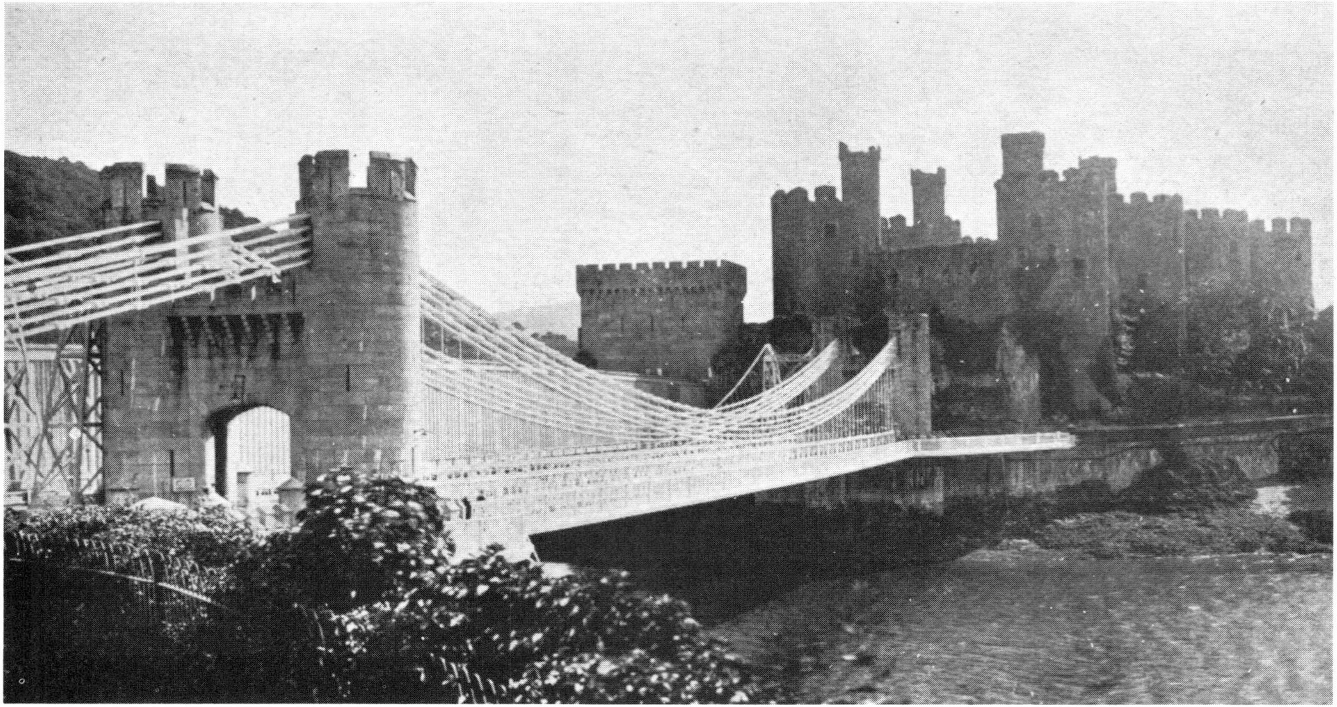

Fig. 53. Conway Castle Bridge, built in 1822–1826 by Thomas Telford

not been finally solved (as has again been emphasized by the collapse, in 1940, of the suspension bridge over the Tacoma Narrows, U.S.A.), engineering science had at the beginning of the nineteenth century, already advanced so far as to permit a largely correct forces analysis of these suspension bridges. It was, in fact, already possible to determine the most economic dimensions of the principal elements, especially the chains or cables, on a scientific basis, in conjunction with strength tests to which the more important among these elements were subjected. In this sphere, it was the pioneer work carried out by Navier which dominated the theory of suspension bridges for a long time. Conversely, the tests and research on building materials, carried out in connection with these great engineering works, served to amplify the knowledge of the properties of these materials in no small degree. For example, the tests carried out by Dufour for the construction of the suspension bridge at Geneva revealed, for the first time, that drawn wires have a greater strength than wires of un-treated or annealed material.[1]

The cast-iron arch and suspension bridges were soon followed by wrought-iron structures, not only in the shape of arches[2] but also in the form of girder bridges for which the new building material was particularly suitable owing to its resistance to compression and tension alike. The earliest important example of an iron girder bridge is the Britannia Bridge across the Menai Straits which was built in 1846–1850 by *Robert Stephenson* (1803–1859), son of the great railway pioneer. The structure consists of tubular girders, assembled from wrought iron plates and angle irons, which bridge the Straits in four spans, two of 460 ft. and two, over land, of 230 ft. each. The masonry piers were carried a considerable distance above the tubular girders so as to permit the strengthening of the girder bridge by suspension cables or chains, if necessary. This precaution, however, proved to be unnecessary. The structure as it stands today, with its rigid vertical and horizontal lines, appears rather formal, somewhat reminiscent of archaic, Egyptian architecture (Fig. 55, facing page 188).

---

[1] cf. "Analyse et extrait des deux ouvrages de M. G. H. Dufour sur les ponts suspendus", in *Annales des ponts et chaussées*, 1832, Vol. 2, page 85.

[2] e.g. the Pont d'Arcole over the Seine in Paris, built by Oudry in 1854-1855.

This novel structure, extremely bold for its time, gave rise to scientific research and, especially, to extensive strength tests which were published, together with a description of the bridge and a history of its construction, in a work of two volumes, written by Stephenson and his collaborator, Clark.[1] With the assistance of *Sir William Fairbairn*, a mechanical and marine engineer who was the most competent expert of the time on the properties of iron, numerous bending tests were carried out with tubular girders of rectangular, circular and elliptic cross-section, and with composite girders of different cross-sections, the mathematician *Hodgkinson* being called in to assist in the evaluation of the results. One series of tests was carried out on a 75 ft. model (corresponding to approximately one-sixth of the length of the longest span) of a tubular girder, the investigations being extended to the influence of live loads and to the buckling problem of the plates.

Soon after the appearance of the first iron bridges, the new material also began to be used in *buildings*, especially for the construction of large-span roofs and domes. Among the buildings erected in this material during the first half of the nineteenth century were the cupola of the Paris Corn Market, with cast iron ribs of 131 ft. span (built by Bellanger, 1809–1811) ; the barrel-shaped roof of the 62 ft. wide indoor swimming pool of the Diana Baths in Vienna (1820) ; the wrought-iron cupola above the east choir of Mayence Cathedral, 49 ft. in diameter (1827) ; the great reading room of the Library of Sainte-Geneviève in Paris (1843–1850) where iron was used not only for the roof structure but also for the supporting columns.

The increasing use of wrought-iron for bridges and buildings is closely related to the development of the mechanical rolling method.[2] This process made it possible to shape the material, in an economic way, into long and thin bars of different cross-sections which could be closely adapted to the purpose and to

---

[1] "The Britannia and Conway Tubular Bridge with general enquiries on Beams and on the Properties of materials used in Construction", Two Vols., London, 1850.

[2] The beginnings of the rolling method go back to the sixteenth century, when the fine metals used for minting purposes were rolled into bars of uniform thickness by means of stretching and cutting mills.

the statical requirements. The production of bar-mill products on an industrial scale had been introduced by Henry Cort in the seventeen-eighties. About 1820, *John Birkinshaw* was granted a patent for the rolling of railway rails, and the first angles were rolled in England around 1830, and in Germany in 1831. Channels and double-tee beams were first mass-produced in France, where a floor-structure carried by 14 centimetres standard double-tee joists was erected in 1849. At about the same time, similar sections were rolled in England and, in 1856, in Germany also.

Of even greater importance to the development which was to render iron the most important material of modern engineering, was an innovation in the production itself. In 1855, the versatile English engineer and inventor, *Henry Bessemer* (1813–1898) conceived the idea of replacing the traditional laborious and costly puddling process by the mechanical process of blowing a blast of air through the fluid pig-iron. The innovation permitted the production of wrought *steel* in large quantities and at really economic prices. Bessemer, a man of all-round ingenuity, has given a vivid description of the epoch-making event in his autobiography.[1]

A further important step forward was the idea which *S. G. Thomas* (1850–1885) conceived in the eighteen-seventies, of reducing the phosphorus contents of the pig-iron by introducing a basic lining into the Bessemer converter, which permits the utilization of the much more common phosphoric iron ores.

These inventions, to which should be added the Siemens-Martin open-hearth furnace with regenerative heating (utilization of waste heat for the pre-heating of the furnace air), opened the way for the extensive use of iron in all fields of technique. Soon, a great number of important engineering works of iron arose in Europe and America, especially bridges, including some of exceptional dimensions which could only be designed and constructed on the basis of exact calculations. The science of structural analysis had suddenly become indispensable to civil engineering.

The subsequent efforts in the field of iron metallurgy were

[1] cf. Matschoss, page 226.

mainly aimed at improving the strength properties of the building material. This is a matter of vital importance to the construction of large bridges. For, any increase of the admissible stresses permits a noticeable reduction of the bar sections and, thus, of the dead weight which, in turn, permits an increase in the length of spans that can be bridged economically. Modern normal-type structural steel, which still does not differ greatly from Bessemer's first steel, has an ultimate tensile strength of approximately 50–60,000 lb. per sq in. But it is now possible to produce structural steels of 100,000 lb. per sq. in. and more by a judicious manipulation of the carbon contents, partly in conjunction with special heat treatment or with small admixtures of nickel, copper, manganese, silicon. Of even greater strength are the wires which form the carrying cables of the great suspension bridges. In the case of the Hudson Bridge of 3,500 ft. span, built in 1927–31, by a native of Schaffhausen (Switzerland), civil engineer, *O. H. Ammann*, a minimum tensile strength of about 230,000 lb. per sq. in. was stipulated for the wires. The bridge over the Golden Gate at San Francisco, which has the world's hitherto greatest central span of 4,200 ft., is suspended on wires with a minimum strength of about 220,000 lb. per sq. in. But most of the modern high-grade special steels, such as nickel-chromium steels and others, which have a strength of up to 260,000 lb. per sq. in., are used for machines, machine tools and weapons rather than for structural purposes.

3. STRUCTURAL ENGINEERING AND "ARCHITECTURE" PART COMPANY

The extensive use of iron and steel as material for bridges and, subsequently, for buildings could not fail to have a profound influence on the development of architecture. The properties of the new material differed greatly from those of the traditional materials, stone and wood. Its tensile and bending strength was considerably higher. On the other hand, the technique of its application was entirely different. It was therefore impracticable to adapt the historical building styles to construction in iron. The building styles of the past were all based on the use of bulky, three-dimensional materials. True, the masters of the Gothic art attempted to reduce the masonry to a minimum, compatible with structural safety. But they, too, made use of profiled vault ribs, capitals, cornices, responds, buttresses and turrets, elaborated from the solid stone, so that the individual building elements have the appearance of tangible, three-dimensional bodies. Iron structures, however, are composed of rolled, mass-produced bars and plates, where the dimension of length, or at most, that of a surface, is predominant. In other words, they give the appearance of being one-dimensional or, at most, two-dimensional.

But, quite apart from this new feature which was, after all, confined to iron construction and need not have affected the methods of masonry construction, there emerged yet another new fact which had, in itself, a decisive and, to some extent, fateful influence on architecture at large. The scientific treatment and solution of the structural problems of statics and strength permitted a more rational design of the structures, and thus made it possible to cope with extensive and difficult structural tasks in an economic way, without prejudice to safety requirements. At the same time, however, it was now possible to design structures according to two points of view, different in principle : the one emphasizing the engineering aspect, i.e., structural analysis and calculation, and the other stressing the architectural aspect, i.e., the æsthetic appearance. All according to the nature of the task, the one or the other of them prevailed : the engineering aspect in the case of utility buildings, and the architectural aspect in the case of monumental buildings. With certain engineering works, such as bridges in towns and the like, both aspects must be reconciled as far as possible.

Up to almost the end of the eighteenth century, the engineer himself had hardly become conscious of the division between engineering and architecture although that division had, to some extent, already taken place. Even books which were specifically addressed to the engineer, such as Bélidor's "Science des Ingénieurs" (cf. page 121), would contain, at least, a brief chapter on "décorations des édifices" and on the classical orders of columns. Engineering specialists also tackled architectural tasks, often with great skill. Gauthey, for instance, built numerous churches.[1] But, with the refinement of the methods of structural analysis during the first decades of the nineteenth century, the scope of theoretical knowledge which the engineer was compelled to acquire through scientific studies, grew to such an extent that specialization became inevitable. Gone were the days of Brunelleschi, Fra Giocondo and Francesco di Giorgio, when technique and mechanics on the one hand, and architecture on the other hand, were equally dependent on three-dimensional imagination. In spite of the civil engineer's predilection for immediately perceptive rather than purely analytical methods, the new branches of science, structural mechanics and statics, were more and more based on theoretical reflections and mathematical deductions which offered little inspiration to persons whose interests and abilities were more inclined towards artistic intuition.

At the outset, the consequences of this split were not vital to the art of building. The rationalist outlook of the later eighteenth century had brought about a revival of classicism. Given good taste and skilful use of classical forms, even the engineer with predominantly mathematical training was able to design buildings which showed at least decorum and good taste, though not necessarily genius. This fact is proved by numerous "classicist" industrial buildings of that time, mills, warehouses, arsenals in all European countries. It is also manifest from many bridge structures where the piers, abutments and arch rings, or the pylons of suspension bridges, offered opportunities for the use of classicist mouldings or arrangements. Perronet's Pont de la Concorde (1787–91) and Rennie's London Bridge are among the finest monuments of classicist bridge building. Rightly, their æsthetic aspect was preserved, as far as possible, when the increasing traffic called for these bridges to be enlarged (in 1930–31 and 1902–04, respectively). From the viewpoint of bridge building history, it is much to be regretted that, contrary to the two bridges just named, the Pont de Neuilly and Waterloo Bridge had to be replaced, just before World War II, by entirely new structures.

---

[1] Another example is provided by U. Grubenmann of Teufen who built churches as well as bridges (cf. J. Killer, "Die Werke der Baumeister Grubenmann". Zürich, 1941).

With the growth of industry and the advent of railways, the need for utility buildings was greatly increased. But whereas, under the "Ancien Régime", engineering works had mainly been carried out for the account of the State, it was now, in the early capitalist era, the private industrialist and the limited liability company for whom the building was erected, and who insisted, first of all, on economic design and construction. Economy became, indeed, the most important criterion of the new era of industrial building, and the theories of structures, as well as of building materials and their strength, had to be made subservient to this purpose.

Apart from housing construction, the architect, working as a creative artist in the traditional sense of the word, was confronted with a marked contraction of his sphere of activities. Even with the tasks still remaining to him, the solution of any unusually complicated statical problems was assigned to the engineer. Small wonder that the architect subscribed more and more to a one-sided, æsthetic approach. In the case of major engineering tasks where the external appearance and, indeed, a certain monumentality was deemed to be of importance, one of the building styles of the past was usually preferred, all the more, as their treasures were now easily available to all those interested, owing to the numerous publications on the history of art. Churches and town halls were built in the Gothic or Romanesque style ; railway stations, post offices and theatres in the style of the Neo-Renaissance. "Buildings of a military character, barracks, arsenals, shooting ranges and the concomitant public houses, or for that matter, any catering establishments with names reminiscent of keeps or castles, were built in the style of pseudo-medieval fortifications."[1] With all these examples, however, it was merely a question of superficially copying the æsthetic, ornamental features of the styles concerned. The inherent motifs, spiritual and structural, which had been at the root of the original styles, had simply ceased to exist.

Any structural problems in connection with, say, the roofing or vaulting of large spaces, the erection of high towers or the

---

[1] Peter Meyer, "Schweizerische Stilkunde" ; Zürich, 1942 ; page 178. The sentence applies, of course, to specifically Swiss conditions.

construction of the first skyscrapers, were solved by the consulting engineer, who applied modern methods and used modern building materials, first iron and steel, and later also reinforced concrete. So as not to prejudice the "style", the structure as such had to be concealed as much as possible. In church naves, the steel or reinforced concrete girders were camouflaged by false, vaulted ceilings. Office buildings with frame structures and light brick walls were faced with a few inches of stone lining, suggesting solid ashlar masonry.

There were only rare exceptions to confirm the rule. In Paris, for instance, some attempts were made, as early as the middle of the century, to reveal the iron structure, designed in accordance with modern principles, and to use it as a decorative element of architecture. The best-known examples are the reading rooms of the Bibliothèque Sainte-Geneviève (1843–50) and the Bibliothèque Nationale (completed in 1861 ; cf. Fig. 56), both designed by *Henri Labrouste*. The last-named structure, in particular, represents, in spite of controversial issues, a very interesting attempt. A. G. Meyer describes it as "the first artistic monument of that new victory over gravity which only iron was able to achieve". Of epoch-making importance in this connection was also London's Crystal Palace (1851), designed by *J. Paxton* (1801–1865).

Perhaps the most successful example of a steel structure which was built not merely for utility purposes, but also for the sake of its architectural appearance, was the Eiffel Tower, erected on the occasion of the Paris Exhibition of 1889. Its designer and constructor, the Swiss engineer *M. Koechlin* (1856–1946), who was in charge of the Design Department of the Eiffel Company, was a pupil of Culmann. He displayed a masterly ability to find, both in detail and for the structure as a whole, that form which was most suited for the novel building material. His work is a worthy companion to the other building monuments of the French capital, and its slender silhouette has become an integral part of the town's skyline. Nevertheless, the case of the Eiffel Tower must be regarded as an exception which will not easily recur, and which does not finally solve the problem of the use of iron for building structures which have to conform to artistic or monumental requirements.

Much less controversial are those engineering works of the nineteenth century which were erected as utility buildings pure and simple, without æsthetic pretensions and without the assistance of an architect. Among them are iron and stone bridges ; dams and weirs ; large halls for industrial or traffic purposes such as the platform hall of St. Pancras Station, London (built by Barlow, 1866–1868), with its 240 ft. span,[1] or the famous "Galerie des Machines" at the Paris Exhibition of 1889, designed by Cottaucin and Ducert, with a span of 363 ft. ; furthermore, buildings for specific purposes, such as drilling towers, cooling towers, gasometers, etc. Such structures may occasionally appear alien to their rural or urban surroundings. On the whole, however, their effect is æsthetically indifferent, neither attractive nor ugly, just because they have no architectural aspirations. They may to some extent be compared, though on a different scale, with those unpretentious rural and forthright utility buildings like mills, timber bridges, sheds, etc. The beneficial effect of an unpretentious utility building pure and simple may, for instance, be appreciated (as happened to the author) in the station precincts of certain North Italian towns where the aspect of a modest locomotive shed or water tower, in the midst of the pompous Jugendstil façades of apartment houses and office buildings, is a downright relaxation for the eye.

Perhaps the most impressive demonstration of the split between "civil engineering" and "architecture" which has now been largely overcome but which was characteristic for that time, can be seen in some of the great railway stations, urban bridges and other public buildings of the second half of the nineteenth century where a certain degree of "monumentality" was stipulated. As far as station buildings are concerned, that at Milan, though only built during the inter-war years of the present century, may almost be regarded as a grandiose corollary of all the sins of the nineteenth century. The clean and functional forms of the platform roofs, erected in iron and

---

[1] In this case, it is only the station roof proper which can be regarded as an engineering work pure and simple ; by way of contrast, the "architectural" façade, clothed in an altogether different "revivalist" style, is rather inconsistent with the honest utility style of the station roof.

glass, which are a fine example of impressive spaciousness and structural elegance, are in sharp contrast to the reception building, where the skeleton of reinforced concrete is hidden behind a clumsy and pompous ashlar "architecture".

Perhaps even more typical for the dualism of contemporary building are (or rather "were", since most of them have fallen victim to World War II) the many pompous river bridges in German towns, erected at the time of the Kaiser, where the portals, designed in the style of awe-inspiring Romanesque fortifications (Rhine bridges at Worms, Mayence, Cologne, and others) or Norman keeps (Norderelbe Bridge at Hamburg) form a strange contrast to the slender, functional silhouette of the iron river spans (Fig. 58, facing page 190).

CHAPTER VIII

CIVIL ENGINEERING DURING THE
NINETEENTH CENTURY

1. HYDRAULIC ENGINEERING – DAMS, TUNNELS, COMPRESSED AIR FOUNDATIONS

In the field of public works engineering, a number of excellent civil engineers were active at the end of the eighteenth and during the first decades of the nineteenth century. Public works were undertaken on a large scale, including extensive river canalization works. Prominent in this particular branch was *Johann Gottfried Tulla* (1770–1828), whose *magnum opus* was the great improvement of the Rhine. At that time, the river, between Basle and Mayence, was in a lamentable state. Numerous river arms with continuously shifting alignment inundated large areas or converted them into swamps. Tulla had studied mathematics, mining and mechanical engineering, and was acquainted with the new discipline of Monge's descriptive geometry which he had learned at the Ecole Polytechnique on the occasion of a trip to Paris. He was engaged on research on hydraulic engineering concerning the velocity of the flow of water in a river bed, fascine-work, the effect of dams, and the like. In 1803, he became Chief Engineer of the Engineering Department of the Electorate of Baden and was entrusted with River Training and Rhine Canalization. Owing to political difficulties, the actual correction and improvement works could not be commenced before 1817. Tulla did not live to see the completion of the works which were completely successful though, according to modern opinion, perhaps somewhat too drastic.

Tulla was also called in to advise on important hydraulic works in Switzerland, including the regulation of the river Linth between the Walensee and the Lake of Zürich, which was carried out in 1807–1816 on the initiative and under the supervision of *Konrad Escher von der Linth* (1767–1823), in accordance with Tulla's plans.[1]

Other engineers concerned with great hydraulic works were : *Carl Friedrich Wiebeking* (1762–1842), who supervised the correction of the neglected Bavarian rivers, including the Upper Danube above Ingolstadt, and published several works on hydraulic engineering.[2]

Woltmann (1757–1837) and *Eytelwein* (1764–1848) were concerned with river training and embankment works in the North German Lowlands. The former was Chief Hydraulic Engineer at Cuxhaven, and since 1812 Director of River and Embankment Works at Hamburg ; the latter Inspector of Dykes at Küstrin and later, Chief Engineer of the Prussian Public Works Administration. Also as technical writers, both have made valuable contributions to practical hydraulics and hydraulic engineering. Woltmann wrote a work of four volumes, entitled "Beiträge zur hydraulischen Architektur" (1791–1799).[3] Eytelwein was the author of "Handbuch der Mechanik fester Körper und der Hydraulik" (1800)[4] and of "Praktische Anweisung zur Wasserbaukunst" (1802–1824).[5]

It is, finally, pertinent to mention *Franz Joseph v. Gerstner* (1756–1832), Chief Hydraulic Engineer of Bohemia, who worked out a project for a ship canal between the Moldau and the Danube, but eventually proposed the construction of a railway as the more advantageous solution. His name is familiar to engineers concerned with harbour works

---

[1] Goethe was profoundly interested in the works planned and carried out by Tulla, especially in the great Rhine Regulation which is said to have inspired the Finale of Faust II where the aged hero turns engineer, seeing the culmination of his life work in the winning of arable land from swamps and lagoons.

[2] Wiebeking is also well known for his numerous beautiful timber bridges, mainly in Bavaria, and for his work "Beyträge zur Brückenbaukunde" ; Munich, 1809 (cf. P. Zucker, "Die Brücke", Berlin, 1921, page 82).

[3] "Contributions to Hydraulic Engineering."

[4] "Manual of Mechanics of Solids and Hydraulics."

[5] "Practical Instructions in Connexion with Hydraulic Engineering."

Fig. 54. "Grand Pont" over the Sarine at Fribourg, built in 1832–1834 by J. Chaley
By courtesy of the Photographic Archives of the Federal Technical College, Zürich

Fig. 55. Britannia Bridge over the Menai Strait, built in 1846-1850 by Robert Stephenson
Fox Photos Ltd.

Fig. 56. Reading Room of the Bibliothèque Nationale, Paris, built in 1861 by H. Labrouste

Fig. 57. Platform roof of Milan Central Station
Photo by courtesy of the Italian State Railways

through the theory of trochoidal wave motion, propounded by him.

The development of hydraulic engineering took a different course from that of building construction, where the introduction of mathematics and mechanics into engineering practice during the second half of the eighteenth century had marked a turning point, and the beginning of an ever-continuing process of rationalization. True, the French engineers of the Ancien Régime, particularly Bélidor, had tried, not only to use the theories of statics and strength of materials to determine the dimensions of structures, but also to apply the basic laws of hydro-dynamics to the design and construction of hydraulic engineering works (cf. Chapter V, Section 4). In the further course of events, however, theoretical hydraulics as formulated by mathematicians like Daniel Bernoulli, Euler and others, proved to be of little practical use. Once again, there occurred a split between theory and practice.

Theoretical hydraulics was separated from *practical* hydraulics. The latter was developed through the efforts of practical hydraulic engineers like Prony, Eytelwein, Darcy, Weisbach, Bazin and others, some of whom were also engaged in research work at technical colleges. The science of theoretical hydraulics (hydrodynamics), on the other hand, was transformed, by the physicists, into a highly artistic mathematical edifice. But in the process the scientists were compelled to resort to the unrealistic hypothesis of a frictionless, incompressible "perfect" or "ideal" fluid. The hydraulic practitioners, however, required simple, realistic formulas based on practical observation, and were less concerned with rational, physical deductions. The hydrodynamic formulæ of Eytelwein, Weisbach, Bazin, etc., which are still to be found, under these very names, in the most popular Continental handbooks of civil engineering (e.g. Förster, "Taschenbuch für Bauingenieure", 4th Edition, 1921 ; page 1145), are therefore rather akin to the empiric rules for the design and construction of vaults, etc., which were in use prior to the advent of building statics and which have been discussed in Chapter IV, Section 2, although they are obviously on a higher level as regards resemblance to reality, accuracy of results, etc.

For a considerable time, theoretical and practical hydraulics developed more or less independently of each other.[1] One of those who attempted to place practical hydraulics, i.e., the problems of the flow of water in pipes and channels, on a more scientific basis was the British engineer and physicist, *Reynolds* (1842–1912), who proved the compatibility of seemingly incompatible observation results by introducing the distinction between "streamline flow" and "turbulent flow" and the notion of "critical speed" (1883). In modern times, the development of hydrodynamics has also been furthered, in no small measure, by the advance of aerodynamics in consequence of the development of aeronautics.

The subject of hydrodynamics gained renewed topicality, as far as the civil engineer was concerned, when the era of hydro-electric development set in at the end of the nineteenth and the beginning of the twentieth century. Since it became practicable to transform mechanical energy into electric energy and vice versa, and thus to transmit power over long distances, water power came into its own again, having temporarily lost its importance during the century of the freely movable steam engine. The design of turbines, which is essentially based on modern hydrodynamics, is the task of the mechanical engineer. But the structural tasks in connection with hydro-electric plants offer plenty of opportunity for the civil engineer to apply modern methods of hydraulics, e.g., in calculating the swell curves of low-pressure plants, or in designing pressure galleries or pipes, etc. Of particular importance for the latter task is the theory of the pressure impulse which has been put forth by *Allievi* and others, and which takes the elasticity of the water into account.

Likewise in connection with the advent of hydro-electric power plants, the construction of *barrages* and *dams* has become a matter of tremendous importance in modern times (especially since the beginning of the twentieth century). The first dams designed in accordance with modern scientific principles were built in France during the second half of the nineteenth century, initially for water supply purposes only. *De Sazilly* (1853) and

---

[1] cf. the Preface to De Marchi, "Idraulica". Milan, 1930.

Fig. 58. Rhine Bridge at Worms (1897)   *By courtesy of the*
                    Photographic Archives of the Federal Technical College, Zürich

Fig. 59. Old bridge over the Vistula at Dirschau (1850–1857)
From Mehrtens, *Vorlesung über Ingenieurwissenschaften*, Berlin, 1908

Fig. 60. Railway bridge over the Firth of Forth (1883–1890)

Fox Photos Ltd.

Delocre (1866) were the first to emphasize that the question of determining the dimensions of a gravity dam should not be regarded merely as a problem of statics (safety against tilting and slipping), but also as one of the strength of the material, so that it was necessary to calculate the *internal stresses* of the dam and to keep them within appropriate limits.[1] The cross-section of the dam was to be determined as that of a body of uniform resistance, having regard to both the extreme cases of the reservoir being full and empty, respectively, and assuming a straight-lined, trapezoidal distribution of the stresses over the cross-sections concerned. Delocre also suggested that a dam

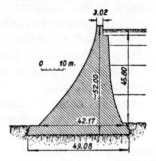

Fig. 61. Cross-section of the Furens Dam (1861–1866). The dimensions are shown in metres.

blocking a narrow valley should be made slightly convex in the plan, and that the vault effect and the resistance of the flanking rock thus invoked should be taken into consideration in determining the dimensions of the lower parts of the dam.

The first major structure designed in accordance with these new principles was the Furens Dam (Barrage du Gouffre d'Enfer), built by *Graeff* and *Delocre* in 1861–1866 for the drinking water supply of Saint-Etienne. The cross-section of this dam (Fig. 61) was known in expert circles as "barrage français" and was, for a long time, regarded as a prototype. The arched crest has a height of approximately 164 ft. and a

---

[1] De Sazilly, "Sur un type de profil d'égale resistance proposé pour les murs des réservoirs d'eau", *Annales des ponts et chaussées*, 1853, Vol. 2, page 191. —Delocre, "Sur la forme du profil à adopter pour les grands barrages en maçonnerie", *Annales des ponts et chaussées*, 1866, Vol. 2, page 212.

developed length of 325 ft. The dam consists of carefully built-up quarry stone masonry, the faces having received special attention.

Delocre's calculation method was subsequently developed and refined in many ways. For the sake of greater structural simplicity, however, the curved cross-section of the Furens Dam was, at most of the later dams, replaced by simpler, approximately triangular cross-sections (e.g. Schräh-Dam, Wäggital, built in 1922–1924). Generally speaking, the subsequent innovations were mostly concerned with details of construction, the aim being to eliminate or neutralize, as far as possible, the influence of the upthrust and of the stress caused by temperature fluctuations and shrinkage. Hundreds of great gravity dams have been built since the turn of the century, including the famous Hoover Dam which has transformed the Colorado River into a lake of nearly 600 ft. depth and 30,000 million cubic yard storage capacity.

In addition to the weight or gravity dam, it was the *arch* dam which assumed an increasing importance during the era of hydro-electric installations, as it was often possible, in narrow gorges, to construct the dam as a horizontal vault and to transmit the vault thrust into the lateral rock abutments. One of the first examples of this type was the Zola Dam near Aix-en-Provence, built about 1850, which is over 120 ft. high.

The calculation of arch dams is among the most important problems of modern structural analysis which has been, and still is, the subject of numerous works and treatises in technical literature. To begin with, one was content to calculate the individual vault rings as fixed arches which had to withstand the horizontal water pressure. In 1913, *H. Ritter*[1] suggested that such a dam should be regarded as a kind of fabric, consisting of vertical beams, fixed at the bottom, and horizontal arches. For the purposes of structural analysis, he proposed to assume such a distribution of the total water pressure over the two statical systems that, in principle, the elastic deflection at all points of the dam remained the same for both systems. As is always the case with the methods of

[1] H. Ritter, "Die Berechnung von bogenförmigen Staumauern". Karlsruhe, 1913.

structural analysis, the practical calculations must be based on greatly simplified assumptions which tend to offset, to some extent, the value of the principle, correct though it is.

During the first decades of this century, many of such arch dams have been built, mainly in America. Swiss examples are Montsalvens (designed by A. Stucky and built in 1918–1921), Amsteg/Pfaffensprung (1921), and others.

The more recent types of barrages (buttress type, etc.) belong to the twentieth century[1] and are thus outside the scope of this chapter. But we still have to deal, briefly, with another branch of civil engineering which has been developed to a high degree of perfection during the nineteenth century, namely, *tunnel construction*. In particular, the great Alpine tunnels built during the second half of the nineteenth century, though hidden in the mountains and, unlike the barrages, concealed from view and without influence on the appearance of the surrounding landscape, have assumed an importance in the history of European transport which can hardly be overrated.

Tunnels through stable rock had already been built in the days of antiquity, and again, on a major scale, since the end of the seventeenth century. In mines and for water supply purposes, galleries of small cross-section have, of old, also been driven through less stable soil (cf. the draining of Lake Fucino, page 2). That the builders of the Renaissance were occasionally also concerned with tunnel construction, is evident from a passage in Filarete's "Trattato dell'architettura" (1464), dealing with the water supply system for the fictitious town of Sforzinda : " . . . The tunnel through the mountain had to be made four miles long. I designed it to be 10 ells wide and 15 ells high, roughly equal to the size of the city-gate. We had to work under great difficulties, encountering rock as well as sand and, in the middle of the tunnel, a heavy black substance . . ."[2]

Dating back to the seventeenth century is the tunnel of the

---

[1] Although a buttress-type dam has, in fact, been built as early as about 1800, in the vicinity of Hyderabad in India ; this dam had a length of half a mile and a height of 40 ft. (cf. Kelen, "Die Staumauern". Berlin, 1926, page 209).

[2] Antonio Filarete, "Trattato dell'architettura", published by W. v. Oettingen in "Quellenschriften für Kunstgeschichte und Kunsttechnik", New Sequence, Vol. 3. Vienna 1896, page 517.

Canal du Languedoc (1679–1681), which has a length of 515 ft., a width of 22 ft. 8 in., and a height of 27 ft. 7 in. The Urnerloch Tunnel in the Schöllenen Gorge was built at the beginning of the eighteenth century (1707). At the end of that century followed a number of other tunnels for French canals.

The beginning of a tunnelling technique in its modern sense, however, dates back to 1803, when the tunnel for the St. Quentin Canal at Tronquoi was driven through layers of sand, i.e., through pressure-exerting material.[1] Other remarkable tunnelling works followed in France and, especially, in Britain for the passage of canals and roads. Among the road tunnels was the Thames Tunnel between Wapping and Rotherhithe, built in 1825–1841 by means of the shield tunnelling method, and later converted into a railway tunnel. This engineering feat is described in "Handbuch der Ingenieurwissenschaften"[1] as a "shining and inspiring example of human perseverance and capability having successfully overcome extraordinary and previously unknown difficulties", and as a work "which places its creator, *Isambard Brunel*,[2] among the most eminent men of his profession for all times. Eleven times the river broke in. But every time the indefatigable engineer started afresh and completed the work after more than sixteen years of strenuous labour".

The advent and growth of the railways gave a mighty impulse to this branch of civil engineering, and the number of tunnels increased to such an extent "that they were soon no longer regarded as technical wonders, but simply as routine tasks of engineering".[1] The first railway tunnels were the Edge Hill Tunnel, near Liverpool (cf. page 169), and a minor tunnel, also on the Liverpool–Manchester Railway, both built under George Stephenson's direction.

---

[1] "Handbuch der Ingenieurwissenschaften", Vol. 5, "Tunnelbau". Leipzig, 1920. Historical Introduction.

[2] Marc Isambard Brunel (1769-1849), a native of Normandy, was architect, civil engineer and mechanical engineer. As an architect, he built the Bowery Theatre and other major buildings in New York. As a civil engineer, he was concerned with the construction of canals and bridges. As a mechanical engineer, he designed, *inter alia*, new types of metal working and textile machines. He was the father of I. K. Brunel, the railway engineer mentioned on page 170 who assisted his father in the construction of the Thames Tunnel.

Up to the middle of the century, the art of tunnelling was mainly based on practical experience, and was therefore the prerogative of a comparatively small number of mining experts. The credit for having founded the *science* of tunnel construction is mainly due to the Austrian engineer, *Franz von Rziha* (1831–1897), who was the first to present, in his classical "Lehrbuch über die gesamte Tunnelbaukunst" (1863–1871), this important branch of civil engineering, with all the relevant types of work and appliances, in a systematic fashion.

As the average length of the tunnels increased, it became a matter of importance to speed up their construction, which is, in turn, dependent on the introduction and perfection of drilling machines. It still took fourteen years (1857–1871) to build the first of the three great Alpine tunnels of the nineteenth century, the eight-miles-long Mont Cenis Tunnel, where Sommeiller's pneumatic drills were used for the first time. Already with the St. Gotthard Tunnel, commenced in 1872, the greater efficiency of Brandt's and Ferroux's drills, and the substitution of dynamite for black powder, permitted a much more rapid progress, so that the $9\frac{1}{3}$ miles tunnel could be completed within nine years. Finally, the first of the two single-track tunnels through the Simplon, which was begun in 1898 and has a length of $12\frac{1}{4}$ miles, could be completed in an even shorter time (7 years, 9 months), although the technical difficulties were considerably greater than at the Gotthard.

Our brief survey of the development of civil engineering during the nineteenth century may be concluded with a few lines on the method of *compressed air foundations*. The use of the diving bell for under-water works has been known for centuries (cf. Coulomb's proposal, page 151). For structural purposes it was first used by Smeaton.[1] In 1841, the French mining engineer, Triger, developed a method of sinking mine shafts which resembles, in principle, the modern methods of compressed air foundation. Triger designed a sheet iron cylinder, which was open at the bottom, closed on top, and filled with compressed air so that the men, working inside,

[1] cf. pages 134 and 170.

were protected against the intrusion of water even when the cylinder was sunk below the ground water level.

In 1843, the English engineer, *Pott*, obtained a patent for a foundation method according to which cast-iron cylinders, fitted with a top lid, were lowered with the aid of the atmospheric air pressure. This was done by producing a momentary under-pressure inside the cylinder, through connecting it suddenly with a large vacuum vessel, so that the external air pressure, acting on the lid, pressed the cylinder into the soil. The method was used for the construction of several bridges in Britain, thus in 1849–1850 by *John Wright*, for the foundation of the Medway Bridge, Rochester. In this case, however, the method was not successful, so that it was replaced by Triger's compressed air method.

The new method marked the beginning of a new epoch in foundation technique. From the very beginning, air locks were resorted to. Two of them were used for each working chamber so that the spoil could be removed without interruption. I. K. Brunel, the railway engineer and bridge builder already referred to (page 170), used the new method for the construction of the Wye Bridge at Chepstow (1850-1856), and the great Saltash Bridge (1853–1856), where a cylinder of 36 ft. diameter, floated to the site, was used for the foundation of the central pier, which was built on a rock 80 ft below high water. The French engineer, *Cézanne*, was the first to use the compressed air method on the Continent when he built the bridge over the River Tisza at Szeged (1857)[1] and the bridge over the Nyemen at Kaunas (about 1859).

The examples mentioned cannot yet be regarded as caisson foundations proper, but rather as tube foundations where the piers consisted, up to a height just above water level, of annular iron cylinders which were filled with concrete or masonry. In some cases, e.g., at the Nyemen Bridge, the lower part was partitioned off as a working chamber and connected by shaft

---

[1] cf. *Annales des ponts et chaussées*, 1859, Vol. 1, page 241. Here Cézanne describes, rather dramatically, the difficulties of the work, and the physical discomforts then suffered by those entering, and staying in, the air locks : buzzing in the ears, earaches, complete darkness, damp and stuffy air, unbearable heat of up to 140° F. and more.

cylinders with the lock chamber, which was located above water level, so that new rings could be added without interrupting the operation of the air locks. The first caisson foundation in the modern sense where the final masonry is placed direct on the top of the caisson and built up above the water level as the caisson is lowered, was used for the construction of the Rhine Bridge at Kehl in 1859.

This was soon followed by a great number of compressed air foundations. The first caissons consisted exclusively of steel. Later on, reinforced concrete and, in North America, timber were also used. Notable examples are the colossal wooden caisson of 102 × 171 ft., used by *W. A. Roebling* for one of the piers of Brooklyn Bridge across the East River, New York (1870–1871), and the steel caisson of 472 ft. length and 135 ft. width, used by *Hersent* for the construction of the dry dock at Toulon (1878–1880).

2. ORIGIN OF GRAPHIC STATICS — APPLICATION TO STEEL TRUSS BRIDGES

After the advent, during the first half of the nineteenth century, of *structural statics* as a separate branch of theoretical mechanics, civil engineering was once more endowed, after the middle of the century, with a novel theoretical aid — *Graphic Statics*.

There were mainly two reasons to cause the advent and development of this new, separate branch of science. One reason was purely practical. The new building material, iron, was being used to an ever increasing extent, chiefly in the form of long, rolled bars, and the particular properties of these building elements favoured the development and increased use of that special form of structure known as *truss*. But the structural analysis of such trusses by means of the analytical methods developed by Navier and his successors was laborious and wearisome.

The other reason had deeper roots. For more than two centuries, the mathematicians had been almost exclusively interested in analytic, arithmetical methods of calculation. But, at the end of the eighteenth and the beginning of the nineteenth centuries, *Monge* (1746–1818) and, particularly, *Poncelet* (1788–1867) made another attempt to shift the emphasis from arithmetical to geometrical methods. The science of *projective geometry*, created by these two Frenchmen, roused the interest of many mathematicians and engineers in geometrical, graphic methods and supplied numerous theorems and designs which were eminently suitable for a clear presentation and elegant solution of problems related to statics and the strength of materials.

In his foreword to the first volume of the "Anwendungen der graphischen Statik" (Zürich, 1888), W. Ritter stresses the importance of projective geometry to the civil engineer with these words : "On the way to graphic statics, projective geometry . . . must be regarded as a natural stepping stone. . . . In the geometrical, graphic treatment of statics, one is continuously confronted with projective and involutary series and families of lines. . . . The tendency to rate Poncelet's wonderful [*sic* !] creations as superfluous, and to seek in a roundabout way what could, with earnest endeavour, be obtained much more easily and naturally by the direct method, is reminiscent of those birds of passage which continue to follow their traditional line of flight and ignore the more favourable ones".[1]

Apart from the graphic composition of forces in Newton's Forces Parallelogram, the origin of graphic statics may be seen in the "Calcul par le trait", used at the Paris "Académie d'architecture" even before the time of the Revolution. The Frenchmen *Poinsot* ("Eléments de statique", 1804) and *Cousinery* ("Calcul par le trait", 1839), as well as the German, *Möbius* (1827), made use of graphic methods for the solution of statical problems, though still without having recourse to Poncelet's projective geometry.

But it was not the French who created graphic statics proper as a separate branch of science. They were scientists working at Swiss, German and Italian Technical Colleges : Culmann and

---

[1] In contrast to this belief, German professors like Mohr, Müller-Breslau or Bauschinger were of the opinion that the study of projective geometry is an unnecessary burden, and that graphic statics could well be taught without it (cf. Rühlmann, pages 475/476).

W. Ritter at Zürich, Mohr at Dresden, Cremona at Milan. The actual founder of the new science was *Karl Culmann* (1821–1881). A native of the Bavarian Palatinate, he had studied at Karlsruhe Technical College and was later, as a young engineer, engaged on railway and bridge building works in Germany. In his spare time, he deepened his knowledge of mathematics and became particularly interested in the new methods of the great French geometricians.

A long journey to England and North America in 1849–1851 had an important bearing on his professional and scientific career. In the United States he had opportunity to study the new truss bridges which were about to be developed at the time, and which were increasingly favoured. Already in 1847, a young mechanic and future bridge builder, *Squire Whipple*, had written "An Essay on Bridge Building", in which he dealt with methods of analysing truss structures. In the report on his journey (published in 1851, in "Försters Allgemeine Bauzeitung"), Culmann was mainly concerned with truss bridges and thereby helped to introduce these structures into Europe.

In 1855, Culmann became Professor of Engineering Sciences at the newly-founded Federal Technical College at Zürich.[1] During his professorial activities there, he wrote numerous publications ; the most important of them, "Graphische Statik", was first published in 1864 and was followed by a second edition in 1875. Departing from a general description of graphic methods of calculation, Culmann develops, in this *magnum opus* of his, the principal graphic methods applicable to problems of structural analysis. In doing so, he amplifies the different individual solutions already known, by numerous new methods and contributions of his own, into a self-contained branch of science, to wit, "graphic statics". In order to obtain a scientific understanding of the principles involved, he is the first to make extensive use of projective geometry, notwithstanding the fact that this method presupposes a fairly high standard of the readers' knowledge. In this respect, Culmann himself remarks (in his Foreword to the Second Edition) that,

---

[1] The College — Eidgenössisches Polytechnikum — was established by a Federal Act, dated 7th February, 1854, and opened in 1855.

if reduced to a limited number of constructional methods, the problems of "statics are deprived of their spiritual essence and thus made more easily digestible to young technicians . . . but that it is necessary for *poly*technical colleges to pursue higher aims . . . the main purpose being to produce thinking human beings". It was not without some difficulty that he succeeded in introducing, at the Federal Technical College, the teaching of advanced geometry which was indispensable for his course on graphic statics.

Culmann is the author of many fundamental graphic methods which still belong to the scientific equipment of every engineer, including the extensive application of the force polygon and funicular polygon, that "principal tool of the graphic structural designer" (Ritter), not only for the determination of bending moments in a beam, but also for the determination of the area moments (static moment, and moment of inertia).

Even in the report on his American journey, Culmann also developed (simultaneously with, but independently of, Schwedler's analogous work, published in "Berliner Zeitschrift für Bauwesen", 1851), a Theory of Trusses[1] which was, however, still based on an arithmetical approach. It was only this theory which "permitted the accurate determination of the dimensions of both flanges and all truss members", and it was "only from that time, when theory and practice began to join hands, that it became possible to design truss structures *economically*."[2] Finally, Culmann enriched the Theory of Earth Pressure by the graphic method named after him (which is, in fact, based on Coulomb's method, but amplified and modified by a systematic substitution of the graph for the calculation).

Culmann's ideas were particularly appreciated in Italy, where *L. Cremona* (1830–1903), with his pamphlet "Le figure reciproche nella statica grafica" (1872), had made a contribution to the theory of plane trusses which proved to be of

[1] Apart from Whipple, already referred to, a Russian engineer by the name of Jourawski had developed, a short while before, a method for the calculation of trusses which remained however, unknown to experts at large. (Timoshenko, "Federhofer-Girkmann Festschrift", Vienna 1950, quoted in *Schweizerische Bauzeitung*, 1951, page 1.)

[2] Heinzerling. "Die Brücken der Gegenwart" ; Aix-la-Chapelle, 1873, Part I, Iron Bridges.

great practical value, namely, the "Stress Diagram", known on the Continent as "Cremona Plan". The author derives his method from projective geometry. But "the method thus derived . . . is so simple that it can easily be understood and applied even without a knowledge of its derivation".[1]

In France, the new methods developed by Culmann and Cremona were made known to a greater circle of experts through a work by M. Lévy, "La statique graphique et ses applications aux constructions" ; Paris, 1874 (second edition, 1886–1888). In 1877, *Williot* developed a method to determine the deflections of trusses, in a treatise on graphic statics.

The science of graphic statics was further amplified and enriched mainly by *Wilhelm Ritter* (1847–1906) and *O. Chr. Mohr* (1835–1918). Ritter was first Culmann's pupil, then (from 1869 to 1873) his assistant, and finally (from 1882) his successor at the Zürich Federal Technical College. It fell to him to edit Culmann's unpublished notes after the latter's death and to expand them into an independent work, entitled "Anwendungen der graphischen Statik" (four volumes, 1888–1906), which covers the entire discipline and which has remained, up to the present day, the classical textbook on graphic statics.

Mohr was a Professor at Dresden Technical College from 1873 to 1900, and does not actually belong to Culmann's school. But engineering science owes him, apart from contributions to the theories of trusses and earth pressure, several important and still popular graphic methods, especially "Mohr's Circle of Inertia", and the construction of the deflection line of a beam by regarding the area of the bending moment diagram as a load and drawing, for this load, a second forces and funicular polygon, with a pole distance equal to E × I. The graphic presentation of stresses afforded by Mohr's Circle is closely related to Mohr's Theory of Ultimate Strength, which is generally appreciated for its particular clarity and very close resemblance to reality. Simultaneously with *Winkler* (1835–1888), though apparently independent of him, Mohr has also introduced into statics the notion and application of "lines of influence" (the name was introduced by Weyrauch).

[1] Ritter, Vol. II, page 8.

The methods of graphic statics are particularly suitable for the calculation of *truss girders* which, as already mentioned, became the most favoured form of structure for large engineering works, especially bridges, during the second half of the nineteenth century. Truss-shaped structures had always been used in timber construction, in the shape of truss and strut frames. In his "Quattro libri dell 'Architettura" (1570), Palladio presents two examples of bridge girders which are, in fact, triangular truss frames. The comparatively rare practical application of this type of structure to wooden bridges may be due to the difficulty of designing the joints in an economic manner.

During the first half of the nineteenth century, wooden truss bridges were much used in North America. Culmann's report on his American journey contains numerous examples of such structures. With the advent of iron, the new building material was used first for tension members (Howe Trusses)[1] and soon afterwards also for entire trusses. On the Continent of Europe, the Belgian engineer *Neville* built, around 1845, some minor canal bridges near Charleroi with an iron truss pattern of equilateral triangles.[2] In England, the first major iron truss bridge was built in 1851 to the design of the engineer, *Warren*, over the river Trent at Newark. In correct appreciation of their statical functions, the compression flanges and struts were made of cast iron and the tension flanges and ties of wrought iron, whilst the joints were hinged. During the following decades, the truss girder became the most popular form of structure in the United States. It was analysed also by other engineers after Whipple, both algebraically and by means of model tests.

Unlike their American colleagues, European engineers preferred the rigid, riveted gusset joints although, to begin with, the assumption of hinged joints was maintained for the sake of analytical simplicity, until, during the eighteen-eighties, Manderla[3] and, subsequently, Winkler and W. Ritter, developed

---

[1] According to W. Ritter ("Der Brückenbau in den Vereinigten Staaten Amerikas", report on a journey, 1893), the aggregate length of Howe's railway bridges consisting of wood with iron ties amounted, at that time, to between one-third and one-half of the aggregate length of all railway bridges of iron.

[2] cf. Heinzerling, as above.

[3] "Die Berechnung der Sekundärspannungen", *Allgemeine Bauzeitung*, 1881.

the exact theory of trusses, taking secondary stresses into account.

In the sphere of the great steel truss bridges, more than in any other sphere of engineering construction, the development of analytical methods and the evolution of statical conceptions was reflected in the change of the external appearance of the structure during the second half of the nineteenth century. During the eighteen-fifties and sixties, a number of girder systems, often named after their inventor, were developed in America, and particularly in Germany, all of them being of a particular, characteristic appearance. One of them was the so-called Pauli Girder (Fig. 62), a lens-shaped truss girder with

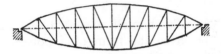

Fig. 62. Pauli Girder.

curved flanges, invented by *F. A. v. Pauli* (1802–1883), an engineer who was born near the town of Worms and was mainly active on railway construction in Bavaria. His idea was to keep the axial forces constant, both in the top and bottom boom, and to obtain a constant length of span, not affected by the load. W. Ritter[1] has this to say about the origin of the Pauli Girder : "At the time of the design, Herr v. Pauli was of the opinion that the failure of bridges was . . . mainly due to the vibrations caused by passing trains. He thought it possible to eliminate, or at least greatly to reduce, these vibrations by suspending the girder in the neutral axis and making this axis a straight line. . . . He was also of the opinion that a variable cross-section of the flanges was undesirable. For, with variable forces in the flanges, there is always a certain waste of material because the cross-section can never be accurately adapted to the forces".

The most important bridge among the numerous Pauli Girder Bridges, mainly to be found in the river basins of

[1] Ritter, Vol. II, page 83.

the Rhine and Danube, was the railway bridge over the Rhine at Mayence, built in 1860–1862.

A few years later (1867), *J. W. Schwedler* (1823–1894) developed a truss system in which all the diagonals were acting as tie members only. If this principle is followed consistently for all bays, one obtains the slight "break" at the centre of the top chord which is characteristic for the Schwedler Girder (Fig. 63). As this form is neither structurally nor æsthetically

Fig. 63. Schwedler Girder.

satisfactory, Schwedler himself suggested that the top chord should rather be given the form of a polycentric or semi-elliptic arch "which has a curvature more pleasing to the eye". "In practice, the Schwedler principle has, in most cases, been followed strictly, and the indeed rather unpleasant form . . . has been preferred".[1]

The Schwedler Girder, too, has been applied to numerous bridges, the first of them being the double-track railway bridge over the Elbe at Hämerten, built by the inventor himself (1867).

The trusses designed by Pauli, Schwedler and others (parabolic girders, semi-parabolic girders, Häseler or K-Girders, etc.) were strut frames which were to be regarded, externally, as simple beams. In 1866, the Bavarian engineer *H. Gerber* (1832–1912) was granted a patent for the design still known as "Gerber Girder", or, in English-speaking countries, as "Cantilever Girder". This is a continuous girder over several spans in which hinged joints are inserted so that the harmful influence of a minor subsidence of supports is eliminated. The first bridge built by Gerber in accordance with this system was that over the River Main at Hassfurt (1867) which was a truss bridge with polygonal chords and a central span of 426 ft. The system can, of course, be applied to any solid-web or truss girder extending over several spans.

[1] Ritter, Vol. II, page 68.

In America and in Britain, the Cantilever Girder found numerous applications, in some cases on a grandiose scale. The first American bridge of this kind was a railway bridge over the Kentucky River built in 1876. In Britain, the great railway bridge over the Firth of Forth was the first and, at the same time, the most magnificent specimen of a cantilever bridge. Built in 1883–1890 to the design of *Benjamin Baker* (1840–1907), the bridge has a length of 8,200 ft. The two centre openings have a clear span of 1,710 ft. each, which represented, for more than twenty years, the longest span in the world (Fig. 60). The cantilever arms are double intersection trusses (see below). The struts consist of tubular steel plates whilst the tie members are, in their turn, formed by braced girders.,

In addition to the (in themselves) statically determinate truss girders with simple or sub-divided triangular pattern, the early part of the second half of the century saw the construction of many structures with redundant members, such as the double intersection truss, with crossed diagonals, and the multiple, narrow-web lattice girder. The difficulty of analyzing the forces distribution in the double strut frame was overcome by designing the diagonals slack, i.e., as tie members only, or by regarding the structure as consisting of two independent strut frames, each taking one half of the load. But the structural analysis of lattice girders was more difficult, so that they were more and more replaced by large-bay trusses. One of the best-known examples of a lattice girder bridge was the old railway bridge over the Vistula at Dirschau (1850–1857), which had six spans of 404 ft. each (Fig. 59, facing page 190).

3. CEMENT AND REINFORCED CONCRETE

If masonry and brick construction were to match the advance of steel construction, it was necessary, first of all, to find a comparatively rapid-setting hydraulic binding agent. True, some Mediterranean countries and other regions of Europe were rich in volcanic deposits, such as pozzuolana ash, found in the vicinity of Pozzuoli (Naples) and near Rome, or trass, found in the German Rhineland, which had been used, since the days of antiquity, for the preparation of hydraulic mortar (cf. page

20). Apart from that, however, the only binding agent used for building construction, up to the end of the eighteenth century, was air-dried mortar, consisting of fat lime and sand. This did not prevent the erection of bold structures, if they were built of accurately cut quarry stone in skilful bond, such as Perronet's bridges. But the reduced strength and, particularly the limited setting ability and slow hardening of air-dried mortar do not permit the construction of highly-stressed structures in cheap quarry stone or cast rubble masonry.

In the case of foundations or hydraulic works, lime mortar was mixed with brick powder, thus enabling the mortar to harden, to some extent, even in water. In Volume IV of his "Architecture hydraulique" (cf. pages 121 and 130), Bélidor describes the manufacture of hydraulic concrete for harbour works, using pozzuolana earth or admixtures of similar effect ("terrasse de Hollande", "cendrée de Tournay").

Towards the end of the eighteenth and at the beginning of the nineteenth centuries, British and French engineers began to take a greater interest in "hydraulic" binding agents, i.e., those hardening in water. In connection with the building of the Eddystone Lighthouse (1756–1759 ; cf. page 170), *Smeaton* carried out experiments with different binding agents and proved that the hydraulic properties of certain species of lime were dependent on their clay contents. For the construction of the lighthouse, however, he used the well-tried pozzuolana earth from Italy. In 1796, *James Parker* was granted a patent for a hydraulic binding agent, obtained by burning natural, calcareous marl with high alumina contents and named, rather misleadingly, "Roman Cement". Around the turn of the century, this product came to play an important part, first in England and subsequently in France, and was soon also used for the preparation of concrete. As early as 1816, a major bridge was built of "Roman Cement" concrete, over the River Dordogne at Souillac, in France.

During the first decades of the nineteenth century, the French engineer, *L. J. Vicat* (1786–1861), carried out detailed, scientific investigations on binding agents. He distinguished between active and feeble fat limes, as well as moderately and strongly hydraulic limes, determining for each type the most

Fig. 64. Reinforced concrete bridge at Wildegg, Switzerland (1890)

By courtesy of Jura-Cement Works, Aarau and Wildegg

Fig. 65. Reinforced concrete stairs at Florence Stadium (1932)

Photo H. Straub

Photo H. Straub

Fig. 66. Gmündertobel Bridge over the River Sitter at Teufen, Switzerland (1908)

favourable degree of combustion. He also envisaged the manufacture of synthetic hydraulic binding agents, burning together pulverized chalk and clay.

But the most important advance on the way to an efficient hydraulic binding agent was made when the English mason and building contractor, *Joseph Aspdin* (1779–1855), succeeded, after long and laborious experiments, in producing the first *artificial* cement, by burning a mixture of clay and chalk (or road dust). This is the wording of Aspdin's famous patent application, dated 21st October, 1824, for the manufacture of "artificial stone" :

"My method of making a cement of artificial stone for stuccoing buildings, waterworks, cisterns or any other purpose to which it may be applicable (and which I call *Portland Cement*) is as follows : I take a specific quantity of limestone, such as that generally used for making or repairing roads, and I take it from the roads after it is reduced to a puddle or powder ; but if I cannot procure a sufficient quantity of the above from the roads, I obtain the limestone itself, and I cause the puddle or powder, or the limestone, as the case may be, to be calcined. I then take a specific quantity of argillacious earth or clay, and mix them with water to a state approaching impalpability, either by manual labour or machinery. After this proceeding I put the above mixture into a slip pan for evaporation either by the heat of the sun or by submitting it to the action of fire or steam conveyed in flues or pipes under or near the pan till the water is entirely evaporated. Then I break the said mixture into suitable lumps, and calcine them in a furnace similar to a lime kiln till the carbonic acid is entirely expelled. The mixture so calcined is to be ground, beat, or rolled to a fine powder, and is then in a fit state for making cement or artificial stone. This powder is to be mixed with a sufficient quantity of water to bring it into the consistency of mortar, and thus applied to the purposes wanted."

Aspdin called this product "Portland Cement" because of its resemblance to Portland Stone which was a very popular building material at the time.

The scientific basis for the manufacture of Portland Cement was provided by Aspdin's compatriot, *J. C. Johnson* (1811–1911) who was, during the eighteen-forties, the manager of a cement factory. In this capacity, he carried out systematic experiments, over a long period, to obtain the best mix of clay and chalk. He recognized the need to continue the burning process until sintering, and invented a number of technical improvements in the production process.

The new binding agent soon became very popular and was increasingly in demand. There also followed soon the first Portland Cement factories on the Continent, 1840 in France (Boulogne) and 1855 in Germany (Züllichau near Stettin).

Well-known are the beginnings of *reinforced concrete*, in as much as they go back to the invention of the Paris gardener, *Joseph Monier* (1823–1906) who, in 1867, took out the first patent for the manufacture of concrete flower pots with a reinforcement of wire meshing. It is less well known that, already several years before Monier, the idea of reinforced concrete construction had occurred to a number of other people and was, indeed, protected by patents. Among these people were the Frenchmen *Lambot* (patent granted in 1855) and *François Coignet* (patent granted in 1861) and, particularly, the American *T. Hyatt* (1816–1901). Lambot had designed, for the Paris Exhibition of 1854, a boat of reinforced concrete which was still in existence at the beginning of this century and may still be in existence today. Coignet was an engineer and intended to produce floors, flat arches, pipes and dams of concrete with steel reinforcement. Already way back in the eighteen-fifties, Hyatt, originally a lawyer by profession, had carried out experiments with reinforced concrete beams which were far ahead of the knowledge of his time, and in which he anticipated certain structural elements that were hailed as new nearly half a century later.[1] In a perfectly correct manner the steel reinforcement of Hyatt's beams was concentrated in the tensile zone, bent upwards near the supports, and anchored in the compression zone by means of vertical steel stirrups. Hyatt laid particular emphasis on the high fire-resistance of the new building element which is partly due to the fact, ascertained by him, that the thermal expansion coefficients of concrete and iron are equal.

But all these early experiments remained more or less unknown, and it is only through Monier that the engineering world at large became acquainted with the new method. Monier's numerous patents,[2] taken out since 1867, relate to

---

[1] Emperger, "Handbuch für Eisenbetonbau", page 14 (see Bibliography).
[2] Some of them are reproduced verbatim in "Handbuch für Eisenbetonbau", page 16.

Fig. 67. Seine Bridge at La Roche-Guyon (1932–1934)

By courtesy of the Archives of the Touring Club de France

Fig. 68. Langwies Viaduct of the Coire-Arosa Railway (1912–1913)

By courtesy of Brown, Boveri et Cie, Baden

containers, floors, beams, pipes, bridges, railway sleepers and other elements.

Monier's first major buildings in reinforced concrete were water reservoirs, including one of 33 cubic yards capacity, built in 1868–1870, and even one of 260 cubic yards capacity, built in 1870–1873. The first reinforced concrete bridge, 52½ ft. long and over 13 ft. wide, was built in 1875.

Owing to Monier's patents, the new method of construction was soon also adopted in Germany, Austria, Britain and Belgium. "It is undoubtedly due to Monier's exceptional energy and practical sense that the conditions were created which prepared the way for the subsequent triumph of reinforced concrete construction."[1]

As far as Monier himself is concerned, his steel reinforcement was mainly designed to obtain the desired shape of the structure and generally to enhance its strength. Statical reflections, let alone calculations to determine the dimensions and shape of the steel reinforcement, were beyond his reach.[2] The first theory for the analysis of composite structures was put forward in 1886 by the German engineer *M. Koenen* (1849–1924). Adopting Navier's assumption of consistently plane cross-sections, he placed the zero line into the central axis of the cross-section and neglected the tensile strength of the concrete.

In 1894, the Frenchmen *Edmond Coignet* (son of François Coignet) and *de Tedesco* submitted a report to the "Société des ingénieurs civils de France" which contained a method of calculation much akin to modern theories. Further vital contributions to the mathematical analysis of reinforced concrete structures are the works of *P. Neumann*, who was the first to prove the relationship conditioned by the elasticity modulus of concrete and steel, respectively ; and a further treatise by Koenen, written in 1902.

Koenen based his theory of reinforced concrete construction

[1] "Handbuch für Eisenbetonbau", page 21.

[2] According to Koenen ("Zur Entwicklungsgeschichte des Eisenbetons", Personal Reminiscences, *Der Bauingenieur*, 1921 ; page 347), Monier once expressed, during a visit to a building site, his disapproval of the allegedly negligent (but, in fact, perfectly correct) manner of execution because the workmen had placed the reinforcement "so far away from the centre of the slab".

largely on the extensive experiments, carried out about 1885 by *G. A. Wayss* in conjunction with "Freytag and Heidschuch" (later "Wayss and Freytag", building contractors) at Neustadt, and in conjunction with "Martenstein and Josseaux" at Offenbach. These experiments, which Koenen watched as an official representative of the Prussian Ministry of Public Works, took the form of strength tests of slabs and vaults to determine the most suitable arrangement of the steel reinforcement. The tests confirmed Koenen's opinion "that it is the main purpose of the steel to absorb the tensile stresses, and that the concrete alone must be expected to absorb compression stresses only. . . . Koenen has thus rediscovered the true statical meaning of the steel reinforcement in the concrete".[1] (As already mentioned, Hyatt's similar tests and perceptions had remained unknown on the Continent of Europe.)

Numerous further tests, carried out by Wayss, Bauschinger (1887), Bach and many others, proved the particularly suitable properties of composite steel and concrete construction also in regard to fire, blow and rust resistance and paved the way to a speedy general acceptance of the new construction method. Among important, early realizations may be mentioned an arch bridge of 131 ft. span and 10 in. crown thickness at the Industries Fair, Bremen, in 1890 ; a road bridge across an industrial canal near Wildegg, in the Swiss Canton of Aargau, with a span of 122 ft., a rise of $11\frac{1}{2}$ ft., and no more than 8 in. crown thickness (1890, Fig. 64);[2] the bridge over the River Ybbs at Gross-Höllenstein, Austria, with a main span of 144 ft. (1896–1897).

In France, reinforced concrete construction was mainly developed by *François Hennebique* (1843–1921). This man of genius designed not only the slabs and vaults, but also the columns and beams of his buildings in reinforced concrete. Of particular merit was his invention of the so-called "slab beam" of T-shaped cross-section in which the economic and statical advantages of the composite construction method are particularly evident. "The basic idea of Hennebique's

---

[1] "Handbuch für Eisenbetonbau", pages 22–23.

[2] Described in *Schweizerische Bauzeitung*, Vol. XVII, No. 11, page 66 (14th March, 1891).

invention is the monolithic conception of his designs . . . his work marks the beginning of a new era of reinforced concrete construction."[1] Among the numerous structures erected by Hennebique's company in France, or by his licensees in other countries (mainly Belgium, Italy and Switzerland) are not only bridges but also, at an early stage, buildings of many kinds (spinning mills, department stores, corn silos, etc.), hydraulic engineering works (quay walls, jetties), ram piles, sheet piling, etc.

It is not possible, within the scope of this treatise, to list all those who have contributed to the further development of reinforced concrete construction. Suffice it to mention some of the most outstanding names. Among the innovations introduced by *Considère* is the so-called "Béton fretté" with spiral reinforcement, permitting the reduction of the cross-section of heavily loaded columns. *Freyssinet* in France and *Emperger* in Austria have performed vital pioneer work and have, *inter alia*, studied the influence of shrinkage, temperature, humidity, etc. *Mörsch* had made an eminent contribution to the development of reinforced concrete construction through his classical textbook, and through the numerous exemplary structures erected by "Wayss and Freytag", the well-known undertaking managed by him. *Melan* introduced the system, often named after him, of rigid, self-stable reinforcement frames on which the concrete forms are suspended ; this system has come to play an important part in the construction of large valley-crossings.

During the early period of reinforced concrete construction, many conventional engineers doubted the feasibility of applying the normal calculation methods, based on the theory of the strength of materials, to reinforced concrete construction, because of the heterogeneous character of the constituent building materials, steel and concrete. These misgivings seemed to be borne out by some major accidents, such as the failure, in 1892, of a 75 ft. test arch at Podol, near Prague, several similar accidents with exhibition buildings in Paris in 1900, or the collapse of the "Hotel zum Goldenen Bären", Basle, on 28th August, 1901. In most countries, special regulations for the calculation of reinforced concrete structures were therefore

---

[1] "Handbuch für Eisenbetonbau", pages 34–35.

introduced by the authorities. The Swiss Institution of Engineers and Architects was the first to introduce, in 1903, "provisional standard specifications" of this kind which were followed, soon afterwards, by similar regulations, drafted by a joint committee of the German Institution of Architects and Engineers and the German Concrete Association. From 1904 onwards, these specifications were regarded as binding by the authorities of many German States. Similar more or less compulsory standard specifications for the design and construction of reinforced concrete structures were officially introduced in France (1906), Italy and Austria (1907), Switzerland (1909) and, during the following years, in most other civilized countries.

In due course, greater confidence was placed in the new construction method. Since the beginning of the twentieth century, reinforced concrete competed to an ever increasing extent, especially in regard to medium-span bridge building, with steel construction which, during the preceding century, had reigned supreme in this field. Among those bridges which must be regarded as milestones in this development, are the following :

The bridge over the River *Isar* at Grünwald[1] with two three-hinged arches of 230 ft. span each (1903–1904), and the *Gmündertobel Bridge* across the River Sitter near Teufen, in the Swiss Canton of Appenzell,[2] with a built-in arch of 260 ft. span (1908, Fig. 66), both designed by Professor E. Mörsch. The calculations, which were based on novel methods developed and published by Mörsch himself, as well as the structural details of these two bridges were, for a long time, regarded as exemplary for many similar structures.

The *Risorgimento Bridge* over the Tiber in Rome, built in 1910–1911 by the Italian subsidiary of the Hennebique Company (Fig. 69). With 100 metres (328 ft.) span and no more than 10 metres (32 ft. 10 in.) rise, this bridge was, at the time, of unprecedented boldness and was particularly suited to reveal the advantages of the new, monolithic building method. The vault, which is no more than 8 in. thick at the crown, is integ-

---

[1] cf. *Schweizerische Bauzeitung*, 1904, Vol. XLIV, Nos. 23 and 24.
[2] cf. *Schweizerische Bauzeitung*, 1909, Vol. LIII, Nos. 7–10.

rated with the superstructure by means of seven longitudinal ribs which are stiffened by cross walls and extended over the abutments so that statically, the whole structure represents a single monolithic, cellular entity.

The *Langwies Viaduct* of the Coire-Arosa Railway (1912–1913 ; Fig. 68), one of the first great railway bridges in reinforced concrete.[1] The great central arch, dissolved into two slightly straddled ribs, has a span of 100 metres (328 ft.) and a rise of 42 metres (138 ft.). The wonderful, slender structure, standing out against the dark background of the woods as it

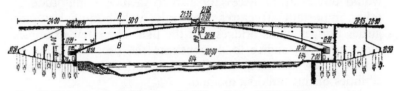

Fig. 69. Risorgimento Bridge over the Tiber, Rome (1910-1911).
From Kersten, *Brücken in Eisenbeton*, Vol. II, Berlin, 1922.

sweeps across the valley, reveals a rare measure of that unity of construction and architectural appearance which we admire in the creations of earlier times, and supplies an impressive proof of the æsthetic possibilities inherent in the new building method.[2]

The *Elorn Bridge* at Plougastel, built by Freyssinet in 1928–1929, with its three main spans of 610 ft. each which represented, at the time, the greatest span ever bridged by a massive structure. The arches, erected in high-grade concrete, have a hollow, box-shaped cross-section. The platform girder is a two-tiered reinforced concrete framework.

The span of the Elorn Bridge was surpassed, ten years later, by the arch of the railway bridge across the *Rio Esla* in Spain[3]

[1] Already several years earlier, two railway bridges with spans of 118 ft. and 197 ft., respectively, had been erected in reinforced concrete on the Kronstadt (Brasov) — Fagaras line, in Transylvania. Experience with these bridges tended to allay earlier misgivings to the effect that the vibrations would endanger the adhesion between steel and concrete (cf. *Schweizerische Bauzeitung* of 29th May, 1909 ; page 287).

[2] *Armierter Beton*, 1915, No. 7 *et seq.*

[3] *Beton und Eisen*, 1935, page 214. — *Génie Civil*, 21st August, 1937, page 161.

which has a theoretical span of 689 ft. and a clear span of 630 ft. This bridge, in turn, was excelled by the *Sandö Bridge* in Sweden, which has a theoretical span of 866 ft.

All the bridges so far mentioned are of the classical design with top platform. Among the most important examples of reinforced concrete bridges with suspended platform is the *Seine Bridge at La Roche-Guyon*, built in 1932–1934 (528 ft. span ; Fig. 67, facing page 208).[1]

In order to convey a reasonably comprehensive picture of the development of reinforced concrete bridge building, it would obviously be necessary also to include a multitude of other important structures, not only arch bridges but also girder bridges, truss bridges and the like. These few examples may, however, suffice to demonstrate the progress, achieved within a few decades, and the structural and æsthetic possibilities of this building method.

As far as *building construction* is concerned, reinforced concrete is extensively employed for columns, beams and slabs, i.e., as the bearing structure which is not discernible from outside. The rich architectural possibilities inherent in the use of ferro-concrete construction have, so far, been exploited to a limited extent only, although the new building method did have an unmistakable influence on the trend of contemporary building styles ("Technical Style", cf. Chapter IX, Section 2). There are, however, certain new structural elements which owe their origin wholly to the advent of ferro-concrete construction and which, due to the particular structural properties of reinforced concrete, are pregnant with novel possibilities of architectural design. They are :

Shell vaults, also known as "concrete-membrane structures", which permit the roofing of large spans with a minimum of material. This method, which has so far mainly been used for large halls, hangars, aircraft sheds, etc., also offers new solutions for structural tasks of a more monumental kind.

Curved and corbelled frameworks which offer an inexhaustible variety of solutions for the design of galleries, flights of

---

[1] *Génie Civil*, 1935, Part I ; pages 125 and 155 (9th and 16th February).

stairs and similar structures (e.g. Florence Stadium, Fig. 65, designed and constructed by P. L. Nervi).

Columns with "mushroom" capitals carrying cross-reinforced slabs without underbeams. This method, which permits the construction of heavily loaded floor structures in a manner equally satisfactory from an economic and æsthetic point of view, goes back to Robert Maillart, whose first experiments in this field took place as early as 1908.[1]

Just because of the almost unlimited adaptability of ferro-concrete construction, it is much more difficult in building construction than in bridge building to achieve that essential unity of structural purposefulness and æsthetic design. Owing to the obvious *raison d'être* of a bridge, the statical function of the structure is all the more clearly visible, the more the length of the span renders it necessary to exploit the strength properties of the building materials to the utmost. There is no doubt, however, that also in other spheres of building construction, reinforced concrete still contains a wealth of untried architectural possibilities.

4. THE MECHANIZATION OF BUILDING TECHNIQUE

In addition to the advent of structural statics and the employment of the new building materials, steel and concrete, the nineteenth century has brought a third revolutionary innovation to civil engineering, and that is the far-reaching mechanization of engineering works. True, the use of elementary machines has always been regarded as a criterion of large-scale engineering works (cf. Chapter IV, Section 3). But, up to the nineteenth century, these machines were almost exclusively confined to hoisting devices which also include, in a wider sense, pile drivers, pumps and the like.

During the eighteenth century, the treadwheel[2] and capstan operated by man or beast, which had so far enjoyed a practically unchallenged monopoly on the building sites, was supplemented by the water wheel.

[1] cf. M. Bill, "Robert Maillart", page 155. Zürich, 1949.

[2] cf. pages 34 and 94 ; also page 126 regarding water wheel drive of scoops and pile drivers, used by Perronet. See also Fig. 35.

In the wake of the Industrial Revolution, the new century brought, apart from the general progress of mechanical engineering due to the use of iron, two innovations of decisive importance : one was the employment of efficient, mobile heat engines for driving purposes, beginning with the steam engine (locomobile engine). The other was the employment of machines for building site purposes other than hoisting, including first of all, mechanical concrete mixers, screening and washing plants, steam and pneumatic hammers, drilling machines, compressors, etc. In addition, enormous improvements were also made on hoisting and transport devices. The old tackles and windlasses and the clumsy cranes with wooden driving gear were replaced by efficient tower cranes, derricks and cable cranes of steel. Locomotive-hauled wagons on temporary iron tracks had many times the capacity of the old horse-drawn vehicles.

Owing to the fact that mechanical engineering in Britain was much ahead of that on the Continent, and was particularly closely interwoven with civil engineering,[1] it was in that country that the mechanization of building plant took its origin. John Rennie the Elder, who had started as a mill designer and machine constructor and became an important civil and mechanical engineer,[2] made history when, during the construction of the London Docks in 1801, he began to use steam engines for the operation of pile drivers, water pumps, etc.

Robert Stephenson was the first to substitute a steam hammer for the old-fashioned ram, during the foundation works for the great railway bridge over the Tyne between Newcastle and Gateshead in 1846. With the aid of the steam hammer, he succeeded in ramming piles of 33 ft. length within four minutes, which permitted a very considerable acceleration of the work.

In the United States, too, the vast scope of the great engineering works soon led to the construction and employment of mechanical devices. Already during the construction of the first Pacific Railway in the eighteen-thirties, steam excavators

[1] cf. page 170 *et seq.*
[2] cf. page 171.

216

were brought into use. In 1858, America produced the first mechanical stone crusher.

Concreting machines were already in use on large building sites around the middle of the century, especially in Germany. Cézanne describes[1] the machine used by him for the construction of the Tisza Bridge at Szeged in 1857 (cf. page 196). This machine consisted of a slightly inclined mixing drum of approximately 13 ft. length and 4 ft. diameter, which was operated by a locomobile engine, by means of a belt drive, the concrete output being approximately 3,000–3,500 cu. ft. per ten-hour shift. A centrifugal pump for site drainage purposes, operated by a locomobile engine, was first used, about 1850, during the construction of the harbour of Oberlahnstein, Germany.

It would be beyond the scope of this treatise to describe in detail how, during the second half of the nineteenth and the first decades of the twentieth century, the design and use of building plant and machinery has developed into a science of its own. As a result of this development, major engineering works now require the assistance of the specialized "plant engineer", in addition to the structural engineer. Great engineering installations for rock or soil excavation and mass conveyance, as well as installations for the preparation, conveyance and placing of concrete for the construction of dams, concrete roads and the like, now require a plant and an installed power which often exceeds that of a medium-size industrial undertaking.

The far-reaching substitution of mechanical work for manual labour has led, in modern times, to a rationalization of the working process which has entailed, not only an enormous acceleration of the work and a considerable reduction in cost, but also frequently an appreciable improvement in quality, e.g., with concrete buildings. In order to extend these benefits also to smaller and smallest building sites, there is nowadays a growing tendency to remove the greatest possible part of the work from the building site to the workshop or factory, where it is possible to resort to industrial production with rationally

---

[1] *Annales des ponts et chaussées*, 1859, Vol. 1, page 241.

designed and permanently installed machinery, instead of the more or less improvised building plant installed on the site. In America, and lately also in certain European countries, even the concrete preparation itself has, in some cases, been removed to the factory where the scientific selection and batching of the materials permits the production of high-quality concrete which is conveyed to the individual building sites in special trucks, equipped with stirring mechanism.

The last few decades have brought important innovations in the following spheres :

In the sphere of *earth movement*, crawler-mounted excavators of a great variety of systems and of the highest efficiency have been developed : grab dredgers, bucket dredgers, diggers. As regards the latter, models of the greatest capacity have been designed, especially in America, e.g. one of 16 cubic yards in 1931. As to wet dredging, the construction of the Panama Canal has given rise to the use of bucket-conveyer dredgers with bucket volumes of up to 2 cubic yards.

In the sphere of *foundation works*, the pile drivers and pneumatic appliances have been further developed and improved, though there have not been any really revolutionary innovations during the last few decades.

As far as *concrete construction* is concerned, it is in the sphere of distributing and placing plant rather than preparing and mixing machinery, that important changes and improvements were made during the first decades of the twentieth century. In America, where the high wages level called for saving in labour, the concrete pouring technique was developed, where plastic concrete is hoisted up to casting towers from which it flows by gravity, through inclined chutes, to the points where it is needed. In Europe, extensive installations of this kind have been used mainly for the construction of the great dams (e.g. Wäggital, Fig. 70). It was found, however, that poured concrete, with its high water content, has certain disadvantages (reduced strength, greater shrinkage), so that recent trends tended to favour cable cranes, conveyer belts and, in certain circumstances, pneumatic pipelines in conjunction with concrete pumps, rather than chutes.

Where high quality is essential, the concrete, placed in the forms, is nowadays condensed by means of mechanical tamping or vibrating devices. For concrete road construction, special roadmaking machinery has been developed.

With regard to the *drive* of building plant, the clumsy steam engine has been almost completely replaced, during the last few decades, by the electric motor or, with mobile plant, by the internal combustion engine.

CHAPTER IX

THE PRESENT

1. RECENT DEVELOPMENTS IN THE SPHERE OF STRUCTURAL ANALYSIS

With the advent of Graphic Statics, the development of Structural Analysis had arrived at a milestone. The main target had been reached. The statical behaviour of the most important structural elements was known, and the engineers were able to determine the design and dimensions of their structures in accordance with scientific principles. The subsequent development of engineering science extended mainly in two directions.

(a) The science of *Building Statics* proper, i.e., the knowledge of the statical behaviour of structures, was further amplified. The aim was to refine the calculation methods and to adapt them more closely to reality. In particular, the methods relating to statically indeterminate structures were systematically improved so as to render them more suitable for practical tasks and less laborious in their application. In addition, new building materials led to the design of new structural systems, calling, in turn, for new methods of calculation.

(b) Parallel with this development, the *theory of the strength of materials*, i.e., the analysis of internal stresses, was also developed further. Here, the main emphasis was on systematical testing, which was also extended to hitherto more or less neglected properties of well-known building materials, and to new materials which had, so far, hardly been explored on a scientific basis at all.

Among the systematic efforts in the sphere of *building statics*, it is pertinent to mention the attempt, made during the

220

Fig. 70. Concreting plant for the Schräh Dam, Wäggital

From *Das Werk*, 1926

Fig. 71. Kill van Kull Bridge, New York (1927–1931) *By courtesy of the Photographic Archives of the Federal Technical College, Zürich*

Fig. 72. Stiffened arch bridge. Bridge of the Rhaetian Railway over the River Landquart at Klosters

Photo S. Berni, Klosters

last decades of the nineteenth century, to reduce the entire science of building statics to a single principle, namely the Principle of Virtual Displacements and *Strain Energy*. As a basic principle of statics, the Energy Theorem was already known for centuries (cf. pages 68 and 112). As far as *elastic* deflections are concerned, the first applications of that principle go back to Clapeyron. *Clerk Maxwell* (1831–1879) and *Mohr* used the principle for the calculation of statically indeterminate trusses. In this connection, Maxwell was the first to prove (1864) the Theorem of Reciprocal Deflection,[1] named after him. "He was the first to determine the relationship between the length alteration of a truss rod and the consequent deflection of a system point, by regarding the truss as a machine through which a driving force P is overcoming a resistance S."[2]

Menabrea (1809–1896) and *Castigliano* (1847–1884) analyzed statically indeterminate systems with the aid of the theorems of the "Minimum Strain Energy" and the "Derivative of Strain Energy". (The first of these two theorems had already been used, in the form of the "Minimum Principle", by Daniel Bernoulli and Euler for the determination of elastic curves, i.e., deflection curves ; cf. page 77).

H. Müller-Breslau (1851–1925) attempted to regard the Energy Theorems as a fundamental principle and to base the entire teaching of advanced building statics consistently on this principle. His work, entitled "Die neueren Methoden der Festigkeitslehre und der Statik der Baukonstruktionen" (first edition published in 1886) has remained the classical textbook in this respect. The calculation methods developed in this work, in which statics is virtually treated as a special case of dynamics, are very much in keeping with engineering mentality, combining as they do, strictness of method with the clarity inherent in the geometrical, mechanical approach. In actual practice, however, the application of these methods to statically indeterminate structures with more than one redundant member

---

[1] "The deflection at a point m in the direction m–m¹, due to a unit load acting in the direction n–n¹, is equal to the deflection at the point n in the direction n–n¹ caused by a unit load acting in the direction m–m¹."

[2] Müller-Breslau, "Die neueren Methoden der Festigkeitslehre und der Statik", Postscript to the 4th Edition, 1913.

leads to equation systems with several unknown quantities, wearisome to solve.

The development of reinforced concrete construction lent special emphasis to one particular aspect of building statics which called for systematic improvements, and that was the analysis and calculation of statically indeterminate *framed structures*. True, it had already been possible to calculate simple frames with the aid of Navier's methods, and more complicated structures such as continuous frames or multi-story building frames by using the energy equations. But the practical process of calculation was, in every case, laborious and wearisome. It was therefore an obvious move to apply the methods which had meanwhile been developed for the continuous beam, particularly the conception of *fixed points*,[1] and to extend them to framed structures. This was done by *A. Strassner* in the first volume of his "Neuere Methoden zur Statik der Rahmentragwerke und der elastischen Bogenträger" (first edition, 1916). The author remarks that "in engineering practice, the methods of calculation based on simple geometrical designs are becoming increasingly popular. They open the way to a natural, perspicuous method of survey and must no doubt be regarded as the basis on which the practical statics of the future will mainly be founded".[2]

The most comprehensive survey of the "fixed-point" methods for the calculation of all kinds of statically indeterminate structures, including multi-story frames and composite frames with straight or curved members in any direction, with constant or variable moment of inertia, has been undertaken by *E. Suter* in his book "Die Methode der Festpunkte", published in 1923. The work, dedicated to the memory of Professor W. Ritter, is an amplification of a thesis written by the same author in 1916 and is, to some extent, based on ideas similar to Strassner's.

Owing to the frequent occurrence of complex frame

---

[1] Points of inflexion of the elastic curve in the unloaded spans of a continuous beam with but one loaded span (introduced by W. Ritter as a further development of Mohr's method of dealing with continuous beams ; cf. Ritter, Vol. III, Foreword and page 24).

[2] Strassner, "Neuere Methoden zur Statik der Rahmentragwerke und der elastischen Bogenträger", Vol. 1, Preface to the Third Edition, 1925.

Fig. 73. Photo-elastic picture

Fig. 74. Housing estate "Weissenhofsiedlung", Stuttgart, designed by Le Corbusier

Fig. 75. St. Antonius Church, Basle   From Reinhardt,
          Die kirchliche Baukunst in der Schweiz, Basle, 1947

structures in building construction, especially reinforced concrete construction, the literature on the statics of such structures has grown immensely in modern times. The method of fixed points has been followed by others, such as the method of angular deflections at the nodal points or the "Moment Distribution Method" by Professor *Hardy Cross*. The principal aim of recent works, written for the practice, is to facilitate the application of these methods and to reduce the time needed for the calculation. For this purpose, they offer ready-made formulas, tables of frequently needed constants, numerical tables, and diagrams. One of these aids, often to be found on the drawing table of the building designer, is the work of a Japanese engineer, *Takabeya*,[1] which permits the rapid calculation of statically indeterminate structures with numerous redundant members, such as multi-span and multi-story rectangular frames as they often occur with multi-floor construction. This book, incidentally, may also serve as an example of a contribution made by a people of non-European race, which has embraced Western civilization only recently.

A similar development has taken place in the sphere of the *built-in arch* which is, like the framed structure, among the structures most frequently encountered in actual practice. This problem, too, can be solved, in principle, with the aid of the energy equations. As early as 1867, Winkler analyzed the fixed-end arch on the basis of the Theory of Elasticity and determined the true position of the thrust line by applying the Principle of Minimum Strain Energy. The fourth volume of W. Ritter's "Anwendungen der graphischen Statik"[2] contains a description of the calculation method, mainly graphic, of the non-articulated arch. This method, developed towards the end of the nineteenth century, is mainly based on "elastic weights" and the "ellipse of stress". What Ritter had mainly in mind was the steel truss arch girder, which was rather popular at the time, a prominent example being the Kirchenfeld Bridge, Berne, built in 1882–1883. The calculation of the left-hand arch of this bridge, which has a span of 81 metres (265 ft. 9 in.), is

[1] Fukuhei Takabeya, Ph. D. "Rahmentafeln". German Edition, Berlin 1930.
[2] Practically completed in 1902 but not published before 1906, shortly before the author's death.

outlined in the book as a numerical example. Another structure which exerted an important influence on the calculation and construction of built-in truss arch girders was the bridge over the River Wupper at Müngsten, built in 1893–97. The grapho-analytical calculation of that bridge, which has a span of 558 ft., has been published in the form of a detailed, printed monograph.

For the calculation of non-articulated concrete vaults, with or without steel reinforcement, the method published in 1906 by Professor *E. Mörsch*, in *Schweizerische Bauzeitung*,[1] was, for some time, regarded as authoritative in the German-speaking countries. The method, which Mörsch used for the design of the famous ferro-concrete arch of the Gmündertobel Bridge near Teufen (cf. page 212), is likewise based on "elastic weights", but is mainly algebraic rather than graphic.

An important step in the direction of rationalization with a view to reducing the mathematical labour of the designer was made by Professor *Max Ritter* (1884–1946) when he published, in 1909, his thesis named "Beiträge zur Theorie und Berechnung der vollwandigen Bogenträger ohne Scheitelgelenk" ("Contributions to the Theory and Calculation of Web Plate Arch Girders Without Crown Hinge"). In Section 2 of his treatise, the author makes certain assumptions regarding the form of the arch and the variability of the cross-section which apply, with reasonable approximation, in most practical cases. He is thus able to anticipate some of the calculations once and for all, and to tabulate the results for the benefit of the reader.

An even more far-reaching reduction in mathematical labour is offered by *A. Strassner* in a method developed in the second volume of the work already referred to. His formulas and tables are likewise based on certain assumptions as to the shape of the arch line and the variation of the cross-section of the vault. They permit an extremely simple statical calculation of the non-articulated vault. By way of example, the author applies the method to the Gmündertobel Bridge and obtains a surprisingly good conformity with the results obtained by

---

[1] Vol. XLVII, page 83 (17th and 24th February, 1906).

Professor Mörsch in the course of his classical method of calculation.

The works of Ritter and Strassner may serve as examples for the many calculation methods of similar purport which enable the modern bridge designer to calculate statically indeterminate, solid vaults on the basis of the Theory of Elasticity, without excessive loss of time.

In modern times, the classical arch bridges and suspension bridges and the different forms of truss girders have been supplemented by a number of other structural systems which

Fig. 76. Vierendeel Girder Bridge over the Scheldt at Avelghem (1904).
From Förster, *Taschenbuch für Bauingenieure*.

have led to the development of appropriate statical calculation methods.

Around the turn of the century and during the first decades of this century, the *"Vierendeel Bridge Girder"* (Fig. 76), named after its inventor, enjoyed a certain popularity, especially in the inventor's own country, Belgium. The Vierendeel Girder is a rectangular frame without diagonals, which was claimed to have certain æsthetic and structural advantages. Its high degree of internal statical indetermination, however, made the calculation rather complicated. As a consequence of several accidents, the use of this form of structure is at present confined to comparatively rare cases where special circumstances call for it.

For large-span through bridges, the use of high-rising, two-hinged or built-in twin arches with *"suspended platform"* is often advantageous. This system has been used for the world's hitherto greatest steel arch bridges, the Kill-van-Kull Bridge in New York[1] with a span of 1,675 ft. between bearing points on the abutments (Fig. 71), built in 1927–1931, and the contemporary Sydney Harbour Bridge of 1,650 ft. span. The Seine Bridge at La Roche–Guyon, a solid-arch reinforced concrete bridge already referred to in the preceding section (page 214), is built to the same principle. In the numerous cases of minor bridges in reinforced concrete, it is often desirable to use the roadway platform as a tension member to absorb the arch thrust so that the bridge can be supported like a simple beam. In recent years, inclined rather than vertical suspenders have sometimes been used for the suspension of the roadway, which offers certain statical advantages. The first bridges of this type were built by the well-known Danish engineering company, Christiani & Nielsen.

We may finally mention, as one of the latest types of bridge, the so-called *"stiffened arch bridge"* which may be regarded, statically, as a kind of inverted suspension bridge with stiffening girder, a well-known example being the bridge of the Rhætian Railway over the river Landquart at Klosters, built by Robert Maillart in 1930 (Fig. 72, facing page 221).

Foremost among novel types of structure in the sphere of *building construction* are the "concrete-membrane structures" and *shell vaults*, already mentioned in connection with the development of reinforced concrete construction. Their calculation, based on principles evolved by Lamé and Clapeyron (1828), has been developed, *inter alia*, by *Dischinger* and *Finsterwalder*, and has been adapted to the practical requirements of the reinforced concrete designer. The calculation is based on the assumption that bending and shearing resistance normal to the shell surface can be neglected, and that the vault, like a sheet of paper, can only absorb stresses acting in the shell surface itself ("Membrane Theory"). On that assumption, the solution can be found with the aid of the equilibrium conditions only, without recourse to elasticity equations.[2] In practice, the calculation method is mainly applied to thin

---

[1] *Engineering News Record*, 1930, Vol. II, page 640.
[2] cf. "Handbuch für Eisenbetonbau", Third Edition, Vol. XII, page 151.

domes and elliptic barrel vaults, often no more than a few inches thick.

In the sphere of *hydraulic engineering*, recent developments in building statics were, and still are, mainly concerned with the stability of dams. H. Ritter's method for the calculation of vault-type dams has already been mentioned (page 192). In this branch of engineering, too, the modern tendency is to facilitate and curtail the work of the designer by means of tables and diagrams (e.g. the works of Guidi,[1] Kelen[2] and many others).

After this short review of recent problems of structural statics, we may turn briefly to the principal developments in recent knowledge of the *strength of materials*. In doing so, we may leave aside the mathematical theories of stresses and elasticity which are, after all, a branch of theoretical mechanics, and thus of physics, and we shall confine ourselves to short notes on those aspects of the subject which are of more immediate interest to the structural engineer and designer.

During the nineteenth century, the dimensions of structures were generally determined so that the "admissible" stress, calculated on the basis of Hooke's Law (cf. page 69), was limited to a certain fraction of the breaking stress, the ratio of the latter and the former being regarded as the "safety factor" of the structure. It was found, however, that the safety factor thus defined does not, in fact, represent a reliable measure of the actual safety of the structure. In the case of many building materials, particularly masonry and concrete, the stresses and strains cease to be proportional even with comparatively modest stresses, as the strains begin to increase more rapidly than the stresses. In consequence, the actual maximum stresses caused by bending are generally smaller than those calculated on the basis of Hooke's and Navier's hypothesis. In the case of steel, it is true, Hooke's Law is sufficiently accurate within the range of stresses admissible in engineering practice. With statically indeterminate structures, however, the local application of a stress beyond the elastic limit will generally lead to a

[1] C. Guidi, "Statica delle dighe per laghi artificiali". Turin, 1921.
[2] N. Kelen, "Die Staumauern". Berlin, 1926.

compensation of forces inasmuch as overstressed parts are relieved by the action of less heavily stressed members. In that case, the real ultimate strength of the structure will, to an extent which varies with the type of structure, exceed the theoretical breaking load calculated on the basis of the theory of elasticity.

This is where the "Theory of Plasticity"[1] intervenes. This theory is concerned with the behaviour of building materials beyond the elastic limit, the ratio of stresses and strains, the effect of fluctuating stresses, the influence of the time factor (called "creeping" in the case of concrete), etc. It is obviously only with the aid of extensive tests that these, often highly complicated, conditions can be clarified.

There are other problems connected with the strength of materials which it is impossible or difficult to solve on the basis of the classical methods of the mathematical theories. Among these problems are the concentrations of stresses, which occur in complex building elements in the immediate vicinity of abrupt changes of cross-section. In this connection, remarkable successes have been achieved with the recently developed "photo-elastic" method of stress analysis. With this experimental method, also known as "stress optics", a small-scale model of the building element concerned (which is regarded as two-dimensional) is built of celluloid, phenolite or similar substances and exposed to polarized light. When the model is subjected to stresses, an appropriate arrangement of the optical apparatus will produce a number of dark lines (isochromatic lines), connecting all points at which the difference of the primary stresses $(\sigma_2 - \sigma_1)$ is identical. Each following line corresponds to two, three, four, etc., times that difference (Fig. 73). From the number and alignment of these lines, it is possible to determine the maximum shearing stress at each point, as well as the magnitude of the edge stresses of the structure.[2]

In addition to this method, other experimental methods are

---

[1] The first theoretical reflections on plastic deformations of solids go back to Saint-Venant.

[2] cf. Föppl and Neuber, "Festigkeitslehre mittels Spannungsoptik". Munich, 1935.

now being used to determine, on a celluloid or similar model, the strains caused by different loads. This is done either optically, with the aid of a microscope, or mechanically with the aid of strain gauges. From these strains, it is possible to draw conclusions regarding the state of internal stresses, and to clarify the behaviour of structures which it is impossible or difficult to analyze mathematically.

The analysis of building structures with the aid of model tests is, in practice, only applicable to special cases. There is, however, another extension of the scientific range of the theory of the strength of materials which is of even greater importance to structural engineering, and that is the exploration of the *building ground* and of that all-important material of deep-level engineering, *soil*. The beginnings of this branch of science, known as "soil mechanics", go back to Coulomb who, in his "theory of earth pressure", already makes use of the important notions of "cohesion" and "internal friction" (cf. page 147). The classical theory of earth thrust was further developed and improved by Poncelet (1835), Rankine (1857), Culmann (1866), Rebhann (1871), Winkler (1872), W. Ritter (1879) and others. Soil mechanics in its modern sense, which is concerned not only with earth pressure on retaining walls but also with settlements, safety of foundations, permeability and stability of earthworks (dams, slopes and the like), is the work of more recent scientists such as *Terzaghi, Casagrande, Fröhlich* and others. This branch of engineering science includes the experimental and mathematical analysis of such soil properties as compressibility, shearing strength, internal friction, permeability and capillarity, as well as the effect of the water pressure in the pores on the plasticity and cohesion of the soil, the influence of the time factor on deformation under pressure, etc.

This outline of certain recent trends affecting building statics and the theory of the strength of materials must, of necessity, remain fragmentary, in view of the immense expansion of the subject matter in modern times. But, as we are dealing not with the history of structural statics, but with that of civil engineering, comprehensiveness is not essential.

For the time being, the practical influence of the new, individual research efforts on the visible shape of building

229

structures is comparatively small. But scientific research work is, after all, the foundation on which progress and development of civil engineering are based. Another factor has a bearing on the subject : in modern times, the share of individual engineers of genius is generally more freely recognized in the field of theoretical research than with the great engineering works. As in the days of Antiquity and in the Middle Ages, the great engineering works are usually due to the anonymous efforts of a great number of collaborators who contribute a large, if not the largest, share in the work. The names of the men who have made prominent contributions to modern civil engineering are usually linked with special *research* objects or *calculation* methods rather than specific buildings or engineering works. True, even scientific research work is more and more becoming a collective effort of whole teams of scientists, and that applies especially to research work in testing laboratories and research institutes, e.g., those for building materials, soil mechanics, hydraulic engineering, and so on. But new methods of calculation and, particularly, the systematic treatment and comprehensive presentation of special branches, are still largely the work of individual personalities.

As regards engineering practice proper, there is some scope for the personal, creative effort of architecturally gifted designers in reinforced concrete construction ; the names of R. Maillart (cf. page 240) or of the Italian engineer, P. L. Nervi (Fig. 65) come to mind. With large-scale buildings, however, and particularly with great engineering works such as dams, canals, harbours or major steel bridges, it is usually difficult to single out the personal contribution of the individuals concerned, as the credit must be shared by the consulting engineers, the authorities in charge and the contractors carrying out the work. Not all the technical reporters are as conscientious as the author of the article on the Kill-van-Kull Bridge in New York, already referred to[1] : he mentions, apart from the Swiss engineer H.O. Ammann who was in charge, the names of no less than nine other engineers who have had a major share in the completion of the great work.

[1] See footnote 1 on page 226.

2. THE INFLUENCE OF ENGINEERING CONSTRUCTION ON MODERN ARCHITECTURE : THE "TECHNICAL STYLE"

In the concluding section of Chapter VII, we have seen how, under the influence of the precipitate development of engineering technique during the second half of the nineteenth century, building construction was approached from two points of view, different in principle. Buildings for industrial or traffic purposes were merely utilitarian in character, designed exclusively in accordance with statical, economic considerations. But where æsthetic considerations did play a part, there was a tendency to over-emphasize the artistic, stylistic aspect.

At the beginning of the twentieth century, the review of recent building history led to a growing awareness of the low architectural standard of modern times. It was found that the "incidental" utilitarian buildings, in contrast to the revivalist creations of contemporary "architecture" did, after all, possess certain artistic qualities which came to be all the more appreciated as people grew more conscious of the hollowness and emptiness of the stuck-on, frontal architecture of the nineteenth century. The sober functionalism of great structures such as corn silos, factory halls, bridges or of machines, vehicles, ships, etc., came to be regarded as an adequate expression of our age, and the shapes which owed their origin to statical, structural and technical considerations began to influence the æsthetic taste as well. Modern engineers endeavoured to make the most of given quantities and properties of material, and to achieve a maximum of efficiency and effect with a minimum of economic effort. The consequent "boldness" of the structure led to a mutual stimulation of technical, structural ability and creative design, and yielded novel types of structures. The process recalls the origin of the Gothic style of vaulting and confirms the opinion that "decisive new forms of style will always originate in those building tasks which are regarded as the most important tasks of their time".[1]

As had happened before in the history of building construc-

[1] Peter Meyer, "Schweizerische Stilkunde", page 194.

tion,[1] a tendency already nascent was furthered, at the decisive moment, by another, incidental, extraneous contingency. The general impoverishment after World War I called for the utmost economy in building construction, for the omission of any superfluous ornamental trappings, and for the utilization of all the structural possibilities offered by modern technique. Like a machine, a structure was to be designed in accordance with the principles of technical purposefulness, and even a house was to be regarded as a "living machine". In France, the protagonist of this school of thought was *Le Corbusier* (Fig. 74). In Germany, where the psychological foundations (viz., the aversion to outmoded trappings) and the material conditions (viz., the need for economic building) were most conducive to the new ideas, it was mainly the team of architects around the "Bauhaus" (Weimar, later Dessau), and among them *Walter Gropius*, who endeavoured to design their housing estates and buildings in accordance with these principles.

Again, history repeated itself. As in the days of the late Gothic, the original structural purpose of the forms tended to be forgotten. With the engineering structures which had given rise to the "technical style", utility and functionalism had indeed been decisive factors. But these considerations tended to recede into the background when the style was adapted to residential and monumental buildings. The rigid rectangularity of the form, the slenderness of the supports, the span of the structures, the projection of balcony slabs un-supported by brackets, all these features were exaggerated beyond technical or economic necessity, and were used as means of æsthetic composition and expression. Those architects who were of real significance were able to combine artistic design with structural purposefulness, so that the term "New Functionalism", often applied to the artistic style of the nineteen-twenties, could justifiably be applied to architecture as well. Their creations embody once more that synthesis of artistic endeavour and structural technique which we have already encountered on several occasions.[2] To many "fellow-travellers", however,

---

[1] E.g., at the transition from Antiquity to early Christian architecture ; cf. page 15.
[2] cf. pages 9, 15, 42.

the flat roofs, the large glazed surfaces, the ostracism of any tradition-bound members such as cornices, pilasters, arches, etc., were merely a matter of fashion. This was immediately obvious to clear-minded critics : "With the 'constructivists', it is obviously seldom a question of real rationalism, contrary to these people's own opinion . . . . Mathematicians, physicists, engineers have little use for this game. We are confronted with the *artistic expression* of the stereometrical, the machinist, the mechanical element."[1]

It would be futile to list examples of structures designed in the "technical style". There are many of them in every major town. Suffice to mention two exemplary church buildings which have had a major influence on the development of the style : Notre Dame du Raincy, near Paris, built in 1922–1923 by *Auguste* and *Gustave Perret*, and the church of S. Antonius at Basle, built in 1926–1927 to the designs of *Karl Moser*. The last-named church has given rise to a popular nickname which aptly epitomizes the engineer-inspired architecture : "Seelensilo" or "Silo of the Souls" (Fig. 75, facing page 223).

The discovery of the artistic qualities of purely technical building forms, and their æsthetic re-appreciation as expressed in the "Technical Style", signifies a turning point in modern architecture which was not to remain without retroactive influence on civil engineering itself. Having been reminded by the artist and architect of the peculiar beauty of his structures, the engineer himself again began to pay more attention to their æsthetic appearance, in addition to their structural and economic aspects. It is a fact that modern engineering literature contains more treatises dealing with æsthetic questions than was previously the case. Even in the sphere of deep-level engineering, for instance with the construction of motorways, attention is being paid not only to traffic considerations but also to æsthetic considerations, such as "landscaping", through a suitable choice of alignment, curvature, slopes of embankments and cuttings, etc.[2]

Building structures such as bridges and dams, buildings

---

[1] Franz Roh, "Nachexpressionismus". Leipzig, 1925. Quoted by H. Sedlmayr, "Verlust der Mitte". Salzburg, 1948.

[2] cf., *inter alia*, "Erfahrungen beim Trassieren der Reichsautobahnen" ; *Schweizerische Bauzeitung*, Vol. 117, page 8 (4th January, 1941).

for industrial or traffic purposes, silos, etc., are often, owing to their dimensions and in many cases also owing to their unusual shape (corn silo, Fig. 77), landmarks which dominate their urban or rural surroundings, and are thus, of necessity, monumental in character. Exceptionality is, in itself, an element of monumentality in as much as it leaves a lasting impression in the mind of the spectator. "To be reminded" is, in fact, the very meaning of a "monumentum" (this is why the designers of monumental buildings have always endeavoured to lift their works above the sphere of the conventional, even in their dimensions. On the other hand, owing to the technical difficulties inherent in such exceptional dimensions, such structures as the Pyramids, the Gothic Cathedrals or S. Peter's Dome must also be regarded as engineering works).

Here, we touch upon a subject matter which has a vital bearing on the "Technical Style". Can engineering structures be monumental, and ought they to be so ? How is it possible to avoid undesirable monumentality ? In order to find an answer to these questions, which have been repeatedly discussed in recent years (especially in Switzerland[1]), it is necessary to distinguish between two shades or grades of monumentality which may be termed "unintentional" and "intentional". The former "is brought about just when the purpose rather than the effect is kept in mind".[2] As already mentioned, such monumentality is inherent in the exceptional dimensions, in the expressive power of statically obvious forms, in the rhythmic repetition of simple motifs like arches, buttresses, pillars, etc., and in the absence of any mitigating ornaments. We talk of "intentional" monumentality if, through deliberate weightiness or pretentious symmetry, the building is assigned an importance which, as a utility building, it often does not possess. Unfortunately, there are many examples of such misplaced monumentality; witness the industrial buildings of Peter Behrens and his school which were highly praised in the nineteen-twenties, such as the Turbine Factory of the General Electric Company, Berlin (Fig. 78). This building makes a somewhat pretentious

---

[1] cf., *inter alia*, "Werk" the Swiss monthly review for architecture, free and applied arts, Vol. 25 (1938), page 123 ; Vol. 27 (1940), pages 160 and 189.

[2] Peter Meyer, *Schweizerische Bauzeitung*, of 4th October, 1924.

impression and can be regarded as an example of "Technical Style" as it should *not* be.

Contrary to popular opinion, even industrial and utility buildingsare never *exclusively* the result of mere utility considerations. This fact, which will be further discussed from a statical angle in the following section, is confirmed by well-designed engineering structures as well as by examples of excesses in the "Technical Style". Engineering structures of the past will always reveal the principal stylistic features of the epoch to which they belong. In our age, even steel and concrete buildings often show certain national characteristics although material and calculation methods are the same everywhere. An eye trained to discern finer shades has no difficulty in finding certain recurring traits of character and stylistic peculiarities which extend, in every case, over the *entire* structural pro-creation of a people, from public or sacral buildings to housing and to purely technical utility structures. In pre-war Germany, for instance, the great bridges and viaducts of the Reichsauto-bahnen revealed an unmistakable stylistic kinship with the representative state buildings of the Third Reich. In Switzer-land, some of *Robert Maillart's* reinforced concrete bridges (Fig. 79, also Fig. 72), with their unpretentious, light-hearted tenor, have certain distinct traits in common with some con-temporary Swiss buildings, e.g., those of the National Exhibi-tion of 1939. Similar phenomena can be discerned in the United States and in other countries.

In the case of certain public engineering works, such as bridges, railway stations, market halls and the like, it is not so much for the designing engineer or architect to decide on the degree of "monumentality". This decision will generally rest with the public authorities concerned. In that case, it may well happen that an engineering structure is deliberately used as a symbol of national might or civic pride, and thus endowed with a certain measure of "intentional monumentality". To express an opinion on the propriety of this conception would be outside the scope of the present work. It may, however, be recalled that similar considerations have also played a part in the past, e.g., with the fortification works of the Middle Ages (cf. page 49), or the Renaissance (cf. page 85).

3. RETROSPECT – THE LIMITATIONS OF THE ANALYTICAL APPROACH TO STRUCTURAL ENGINEERING[1]

The development of civil engineering from ancient beginnings to the present day, outlined in the preceding chapters, may perhaps be summed up as follows :

The Ancient World saw great engineering works, some of them of gigantic dimensions ; but there was no trace of any distinction, in principle, between architecture and civil engineering, nor of any application of exact scientific methods to engineering construction. The same can be said of the Middle Ages.

The Renaissance is the curtain raiser to natural science in its modern sense. The inspired, open minded thinkers of that age, particularly in Italy, break the gates of medieval narrow-mindedness wide open. They seek universality, and they attempt to bridge the gap between art and science, between traditional craftsmanship and theoretical cognition. But the attempt is short-lived and leaves little trace in subsequent history.

The growing scope of science calls for specialization, and research becomes the domain of physicists and mathematicians. Engineering practice and engineering science continue to follow a different course.

The middle of the eighteenth century brings the first attempt to apply scientific methods of structural analysis to stability surveys of buildings, and to the determination of structural dimensions. Within less than a century, the theory of structural statics has been developed, enabling the engineer to predict, in mathematical terms, the statical behaviour and safety of structures, and thus to determine their dimensions and shape on the basis of economic considerations.

During the nineteenth century, the development of scientific methods of calculation, and the discovery and commercial production of efficient new building materials, coincide with the advent of untold new building tasks, arising in the wake of the turbulent development of industry and world-wide traffic.

---

[1] Parts of this concluding section have been extracted from an article by the author, published in *Schweizerische Bauzeitung*, Vol. 116, page 239.

Fig. 77. Corn silo, Chicago   From *Schweizerische Bauzeitung, Vol.* 87

Fig. 78. Turbine factory of the General Electric Company ("A.E.G."), Berlin, built by Peter Behrens, Berlin, 1909

From *Bauten der Arbeit*, Leipzig, 1929

Fig. 79. Reinforced concrete bridge over the River Arve, Geneva, built by Robert Maillart

From *Das Werk*, 1940

Each decade tends to outstrip its predecessor by greater and bolder structures, so that the nineteenth century may truly be regarded as the "heroic age" of civil engineering.

At the present time, the subject matter encompassed under the headings "technical physics", "statics", "theory of elasticity", "theory of the strength of materials", "testing of materials" has expanded to such an extent that it is impossible for the average engineer to master it all. The trend is, therefore, again towards a new separation of scientist and practician (and, of course, towards a further specialization within each of the two branches). In this process, there is a distinct tendency to make the results of research and science available in such a form that the practising engineer is able to use them in the most economic manner. In considering the recent development of building statics,[1] we have seen how, for the analysis of frequently used structures such as frames, fixed arches and the like, methods have been developed which permit the use of numerical tables for the rapid calculation of such statically indeterminate structures. Even further in this direction go the many collections of "ready-made" formulas and tables, available on the book market, or contained in popular handbooks and textbooks, or dispersed in technical periodicals. Examples are Kleinlogel's "Frame Structure Formulas" ; Griot's or Winkler's "Tables of Moments" ; Krey's "Tables of Earth Pressure" ; juxtapositions of moments of inertia and resistance ; numerical tables and diagrams to calculate the stresses and necessary dimensions of reinforced concrete cross-sections, etc. In the shape of nomography, modern mathematical science has endowed the engineer with an efficient and elegant means to present even comparatively complicated relationships between three or more variable quantities in a lucid, diagrammatic manner so that, if two quantities are given, the third and possibly further quantities can be read off without difficulty.

In contrast to scientific research work, the purpose of such publications is essentially formal or systematic, namely to present problems already solved in principle, in a concise and, to the practician, convenient manner so as to curtail the

[1] Page 220 *et seq.*

laborious calculation work with pen and slide rule. The relief thus afforded is meant not only to enhance the efficiency of the practical engineer but also to enable him to devote more time to other, more personal tasks such as organization, planning, creative design or construction.

However, if the use of such aids is not to be prejudicial to the safety or economy of the structures, the engineer using them must have a firm knowledge of their limitations, a highly developed statical feeling and a thorough acquaintance with the properties of the building materials concerned. Such knowledge is acquired and deepened, partly through the practice, and to an even greater extent, by thorough theoretical studies. It therefore goes almost without saying that, for the academically trained technician, theoretical studies and scientific methods can never be *replaced* by collections of formulas or tables. The importance of the latter lies in a rather different plane : they are likely to modify the relationship of the two kinds of work in the sense that the purpose of the theoretical studies will, to a greater extent than before, be *educational* rather than practical.

Even though the formulas and tables, based on simplifying assumptions, will only permit the analysis of the statical behaviour of a structure with a more or less crude approximation, the accuracy to be obtained is, in most cases, perfectly adequate for all practical purposes, even with comparatively important structures. At the risk of repeating the obvious, this fact must be emphasized again and again : The *uncertainties in the basic constants* (ultimate strength, modulus of elasticity, elastic limit, thermal expansion coefficient, degree of shrinkage) are so great that any seemingly greater accuracy of results obtained through refined methods of calculation or through consideration of all incidental contingencies is, in reality, illusory. This is quite apart from the uncertainties inherent in the effects of the construction and concreting process, the pliability of the soil, and the often arbitrary load assumptions. Every expert knows how great these uncertainties may be, especially with masonry and concrete structures. Owing to the summation (or rather multiplication) of the individual errors, the final result may therefore be affected by a fairly great error. It was found, for instance, that reinforced concrete structures which were subjected to breaking tests, showed a degree of safety many times greater than could have been expected from the calculations. On the other hand, an unfortunate coincidence of adverse circumstances and unsuitable design can lead to the contrary case. This is, unfortunately, proved by numerous failures in the history of civil engineering. The cautious and responsible designer will therefore insist on choosing a "safety coefficient" (i.e., a ratio of breaking stress and admissible stress) which is none too scant.[1]

In the case of iron and steel structures, the results of calculation are more reliable in as much as the ultimate strength, modulus of elasticity and thermal expansion of metals can be far more accurately predicted than those of concrete. Even here, however, the uncertainty of the load assumptions remains, especially with regard to wind pressure, seismic forces, etc.

In any case, the effort required to push the theoretical accuracy of calculation beyond certain limits is mostly out of all proportion to its usefulness. It is, therefore, only to be welcomed if the emphasis is shifted from laborious, bewildering and complicated calculation work to overall considerations of good and purposeful design, *in toto* and in detail, i.e., to correct proportions and dimensions and to the integration of the structure with its surroundings.

It is possible that the future will re-establish the greater distinction between the research worker on the one hand, and the creative builder and designer on the other hand, in lieu of the distinction between engineer and architect, established for a century and a half. Certain indications of such a trend have been apparent for some time. In this connection, it is worth recalling the following extract from Professor H. Jenny-Dürst's obituary[2] of *Robert Maillart* (1872–1940), one of the foremost creative civil engineers of Switzerland : "What is the most distinguishing feature of Maillart's structures is their

[1] An exception to this rule applies where the safety of a structure depends on purely *statical* considerations, e.g., in the case of the tilting strength of a block of masonry on which most of the mentioned constants (ultimate strength, modulus of elasticity, thermal expansion coefficient, shrinkage coefficient) have no influence.

[2] *Neue Zürcher Zeitung*, 2nd May, 1940.

independent, *freely creative design*. As far as their statical calculation is concerned, these structures are mainly conceived so as to take account of the primary stresses. . . . Detailed theoretical investigations of secondary stresses were forestalled by suitable structural measures, designed by a man of genius." Maillart himself has expressed his opinion on this subject as follows : "It is admittedly a fairly widespread opinion that the dimensions should be unequivocally and finally determined by calculation. However, in view of the impossibility of taking into account all possible contingencies, any calculation can be nothing but a guidance to the designer."

Maillart's designs are examples of the "Technical Style" *par excellence*. They are homogeneous creations in which shape and structure form a perfect unity. Men of artistic interests have expressed regret "that Maillart never had the opportunity to create buildings of predominantly architectural importance, such as a church".[1] But those works which he was called upon to create, that is to say, mainly bridges, are, though not of unusual dimensions, such as to secure a permanent place for their creator in the history of civil engineering.

For nearly two centuries, civil engineering has undergone an irresistible transition from a traditional craft, concerned with tangible fashioning, towards an abstract science, based on mathematical calculation. Every new result of research in structural analysis and technology of materials signified a more rational design, more economic dimensions, or entirely new structural possibilities. There were no apparent limitations to the possibilities of analytical approach ; there were no apparent problems in building construction which could not be solved by calculation.

In the future, however, the civil engineer concerned with building construction will no doubt again remember that aspect of his work which has perhaps been too much neglected during the past century : the creative design of the structure, its formal and æsthetic perfection, and its adaptation to the circumambient landscape. But such re-orientation will not

[1] M. Hottinger in *Werk*, 1940, page 329.

affect this vital fact : it is the *synthesis* of freely creative design and analytical approach which is the foremost symbol and criterion of civil engineering.

The gradual development of modern civil engineering from its twin origins, traditional building craft and exact science, has been the main subject of this book. To all those interested in the art of building, and to all historically-minded civil engineers, the contemplation of the period in which that synthesis has taken place will always be an object of particular attraction.

	Engineering works Important scientific works and dates relevant to civil engineering
1450	1434–36 Dome of Florence Cathedral vaulted by Brunelleschi
1460	1457–60 Ship canal connecting Milan with the River Adda constructed. ("Naviglio della Martesana")
1470	1461 Old "Nydeck Bridge" at Berne commenced
	1474 Ponte Sisto over the Tiber in Rome reconstructed by Baccio Pontelli
1480	1480 Ponte del Diavolo at Cividale constructed (two arches of 85 ft. span, destroyed in 1917)
1490	Covered timber bridges across the Rhine at Säckingen and Constance
1500	1502 Leonardo da Vinci Military Engineer in the service of Cesare Borgia
1510	1506 Reconstruction of St. Peter's, Rome, commenced by Bramante
1520	1507 Pont de Notre Dame, Paris, built by Fra Giocondo (six arches of 31–57 ft. span)
1530	Old timber bridges replaced by stone bridges on the St. Gotthard Road
1540	1543 Genoa Lighthouse constructed by G. M. Olgiato (height 230 ft.)
1550	1546 "Quesiti et inventioni diverse" by Tartaglia
1560	1559 Ponte Carraia, over the Arno, Florence, reconstructed by Ammannati
1570	1566–69 Ponte di S. Trinità, Florence, reconstructed by Ammannati, (blown up in 1944)
	1577 "Mechanicorum Liber" by Guidobaldo del Monte
1580	Approximately 1580 Gravity dam at Alicante (Spain) constructed
	1586 Obelisk on S. Peter's Piazza, Rome, erected by D. Fontana
	1586 "Mathematicorum Hypomnemata de Statica" by Simon Stevin
1590	1588–90 S. Peter's Dome, Rome, vaulted by Giacomo d. Porta and D. Fontana
	1588–92 Rialto Bridge, Venice, built by Antonio da Ponte
1600	1583–97 Barrel vault of S. Michael's, Munich, constructed (span 66 ft.)
1610	1609 Construction of new harbour basin at Leghorn commenced
	1619 First printed logarithmic tables published by John Napier
1620	Approximately 1625 Rhine — Maas Ship Canal constructed between Rheinberg and Venlo
1630	Approximately 1626 Strength tests carried out by Mersenne
1640	1638 "Discorsi e dimostrazioni matematiche . . ." by Galilei
1650	1653 The "Pont de Berne" timber bridge over the Sarine at Fribourg constructed

	Great scientists and engineers
1450	*Brunelleschi* (1377–1446)
1460	*Leon Battista Alberti* (1404–1472)
1470	*Francesco di Giorgio Martini* (1439–1502)
	Fra Giocondo (approximately 1433–1515)
1480	*Bramante* (1444–1514)
1490	*Giuliano da Sangallo* (1445–1516)
1500	*Leonardo da Vinci* (1452–1519)
1510	*Michelangelo* (1475–1564)
1520	*Tartaglia* (early sixteenth century — approximately 1568)
1530	*Cardano* (1501–1576)
	Francesco de Marchi (1504–1577)
1540	*F. Commandino* (1509–1575)
1550	*Antonio da Ponte* (approximately 1512–1597)
1560	*Benedetti* (1530–1590)
1570	*Giacomo della Porta* (approximately 1540–1602)
1580	*Domenico Fontana* (1543–1607)
	Guidobaldo del Monte (1545–1607)
1590	*Bernardino Baldi* (1553–1617)
1600	*Simon Stevin* (1548–1620)
1610	*Galilei* (1564–1642)
1620	*Mersenne* (1588–1648)
	Descartes (1596–1650)
1630	*Roberval* (1602–1675)
1640	*Mariotte* (1620–1684)
1650	*Hooke* (1635–1703)

	Engineering works Important scientific works and dates relevant to civil engineering
1660	1666 French "Académie des Sciences" founded
1670	Approximately 1673 Leibniz discovers the basic principles of the Infinitesimal Calculus
	1667–81 Canal du Languedoc constructed
1680	1671–83 Dunkirk Port and Fortress constructed by Vauban
	1685–88 Construction of the great water supply duct for the Park of Versailles commenced (not completed)
1690	1687 "Philosophiæ naturalis principia mathematica" by Newton
1700	1691 Hooke's Law
1710	1707 "Urnerloch" on the St. Gotthard Road penetrated
	1707 Tabular juxtaposition of the results of bending tests by Parent
1720	1717 "Principle of Virtual Displacements" formulated by Johann Bernoulli
	1720 "Corps des ingénieurs des ponts et chaussées" founded
1730	1729 "Science des ingénieurs" by Bélidor
1740	1742 Statical survey of S. Peter's Cupola, Rome
	1744 Euler's buckling formula published in "Methodus inveniendi lineas curvas . . ."
1750	1747 "Ecole des ponts et chaussées" founded
1760	1758 Timber bridge over the Rhine at Schaffhausen built by Grubenmann
	1768–74 Pont de Neuilly, over the Seine, built by Perronet
1770	1773 "Essais sur une application des règles . . ." by Coulomb
	1776–79 Cast-iron bridge over the Severn, built by Abraham Darby III
1780	1784 Iron produced by the puddle process, by H. Cort
	1787 First bar iron rolling mill
1790	1786–91 Pont de la Concorde, Paris, built by Perronet
	1794 Paris "Ecole Polytechnique" founded
1800	1796 First chain-suspension bridge in North America constructed by J. Finley
1810	1801 Steam engine used to drive building machinery (rams, pumps)
	1803 First tunnel through pressure-exerting soil constructed (Tronquoi, St. Quentin)

244

	Great scientists and engineers
1660	*Vauban* (1633–1707)
1670	*Carlo Fontana* (1634–1714)
	De la Hire (1640–1718)
1680	*Leibniz* (1646–1716)
1690	*Newton* (1642–1727)
	Jakob Bernoulli (1654–1704)
1700	*Varignon* (1654–1722)
	Parent (1666–1716)
1710	*Johann Bernoulli* (1667–1748)
	Poleni (1685–1761)
1720	*Musschenbroek* (1692–1761)
	Bélidor (1697–1761)
1730	*Daniel Bernoulli* (1700–1782)
	Euler (1707–1783)
1740	*Buffon* (1707–1788)
	Perronet (1708–1794)
1750	*Boscowich* (1711–1787)
	Chézy (1718–1798)
1760	*John Smeaton* (1724–1792)
	Gauthey (1732–1806)
1770	*Coulomb* (1736–1806)
	Abraham Darby III (1750–1791)
1780	*Monge* (1746–1818)
	Franz Joseph v. Gerstner (1756–1832)
1790	*Macadam* (1756–1836)
	Thomas Telford (1757–1834)
1800	*Johann Gottfried Tulla* (1770–1828)
	Woltmann (1757–1837)
	Eytelwein (1764–1848)
1810	*Tredgold* (1788–1829)
	Navier (1785–1836)

	Engineering works Important scientific works and dates relevant to civil engineering
1820	1822–24 First cable bridge in Europe constructed at Geneva by Dufour and Séguin
	1824 First patent for the manufacture of Portland Cement taken out by Aspdin
	1825 Stockton — Darlington Railway opened for public transport
	1826 "Résumé des leçons données à l'Ecole des ponts et chaussées", by Navier
1830	1829 Stephenson's "Rocket" wins the locomotive race at Rainhill
	1831 First angle iron rolled
	1832–34 "Grand Pont" over the Sarine, at Fribourg, built by J. Chaley
1840	1840–44 Nydeck Bridge, Berne, constructed (solid arch of 159 ft. span)
	1846–50 Britannia Bridge over the Menai Strait built by Robert Stephenson
1850	1850 First compressed-air foundation used for the construction of the Medway Bridge, Rochester
	1851 First major steel truss bridge in North America
	1854 Zürich Federal Technical College founded
1860	1855 Bessemer steel produced
	1861–66 Furens Dam built by Græff and Delocre
	1865 "Graphic Statics", by Culmann
	1867 Monier's first patent on reinforced concrete
1870	1869 Suez Canal completed
	1871 Mont Cenis Tunnel completed
1880	1881 St. Gotthard Tunnel completed
	1882–83 Kirchenfeld Bridge, Berne, constructed (two steel spans of 285 ft.)
	1886 "Die neueren Methoden der Festigkeitslehre . . ." by Müller-Breslau
	1886 Method for the statical calculation of reinforced concrete developed by M. Koenen
	1889 Eiffel Tower, in Paris, constructed to the design of M. Koechlin
1890	1890 Railway bridge over the Firth of Forth completed (1,710 ft. span)
	1890 First reinforced concrete bridges built (Bremen Exhibition ; road bridge at Wildegg)
1900	1899–1900 Pont Alexandre III, Paris, constructed (three-hinged steel arch of 353 ft. span, rise/span ratio 1 in 17)

	Great scientists and engineers
1820	*Poisson* (1781–1840)
	George Stephenson (1781–1848)
	Aspdin (1779–1855)
1830	*L. J. Vicat* (1786–1861)
	Cauchy (1789–1857)
	Poncelet (1788–1867)
1840	*H. Dufour* (1787–1875)
	Lamé (1795–1870)
	Clapeyron (1799–1864)
	Robert Stephenson (1803–1859)
1850	*Saint-Venant* (1797–1886)
	F. A. v. Pauli (1802–1883)
	Culmann (1821–1881)
1860	*Maxwell* (1831–1879)
	H. Bessemer (1813–1898)
	J. W. Schwedler (1823–1894)
1870	*Castigliano* (1847–1884)
	Monier (1823–1906)
	Franz v. Rziha (1831–1897)
1880	*Cremona* (1830–1903)
	Mohr (1835–1918)
	Wilhelm Ritter (1847–1906)
1890	*H. Gerber* (1832–1912)
	Reynolds (1842–1912)
	Hennebique (1843–1921)
	Koenen (1849–1924)
1900	*Müller-Breslau* (1851–1925)

247

SELECTED BIBLIOGRAPHY

In numerous cases, the full title of sources has been quoted in the text or in footnotes.

In the case of the following repeatedly quoted books, the references in the text and footnotes have mostly been confined to the name of the author or to an abbreviated title.

Alberti	Leon Battista Alberti, "De re ædificatoria"; Italian Edition, "I dieci libri di architettura". Rome, 1784.
Beck	Ludwig Beck, "Die Geschichte des Eisens in technischer und kulturgeschichtlicher Beziehung", five volumes. Brunswick, 1891–1903.
Bélidor	B. F. de Bélidor, "La science des ingénieurs dans la conduite des travaux de fortification et d'architecture civile". Paris, 1729.
Bélidor	B. F. de Bélidor, "Architecture hydraulique", four volumes. Paris, 1750–1782.
Brunet	P. Brunet and A. Mieli, "Histoire des sciences: Antiquité". Payot, Paris, 1935.
Coulomb	C. A. de Coulomb, "Essai sur une application des règles de maximis et minimis à quelques problèmes de statique relatifs à l'architecture"; reprinted as an Appendix to: Coulomb, "Théorie des machines. . . ." Paris, 1821.
Dehio	Georg Dehio, "Geschichte der Deutschen Kunst", three volumes of text and three volumes of illustrations. Berlin and Leipzig, 1921–1926.
Dehio	Georg Dehio and G. v. Bezold, "Die Kirchliche Baukunst des Abendlandes". Stuttgart, 1892–1901.
Duhem	P. Duhem, "Les origines de la statique", two volumes. Paris, 1905–1906.
Encyclopedia Britannica Ninth Edition, Edinburgh.	
Enciclopedia Italiana	
Feldhaus	F. M. Feldhaus, "Die Technik der Antike und des Mittelalters". Potsdam, 1931.

Galilei	G. Galilei, "Discorsi e Dimostrazioni matematiche intorno a due nuove scienze". Leyden, 1638.
Gauthey	E. M. Gauthey, "Traité de la construction des ponts", published by Navier ; New Edition, 1843.

Handbuch der Architektur, Volume II : "Die Baukunst der Etrusker und Römer," by J. Durm ; Second Edition.

Handbuch für Eisenbetonbau, by Emperger, Third Edition. In the absence of notes to the contrary, the reference relates to Volume I ; Berlin, 1921 ; Chapter I : "Die Grundzüge der geschichtlichen Entwicklung des Eisenbetonbaues," by M. Foerster.

Handbuch der Ingenieurwissenschaften, mainly Volume V, "Tunnelbau". Leipzig, 1920.

Mach	E. Mach, "Die Mechanik in ihrer Entwicklung" ; Ninth Edition. Leipzig, 1933.
Marcolongo	R. Marcolongo, "Leonardo da Vinci, Artista-scienziato". Milan, 1939.
Matschoss	C. Matschoss, "Männer der Technik — Ein biographisches Handbuch". Berlin, 1925.
Matschoss	C. Matschoss, "Grosse Ingenieure", Second Edition. Munich, 1938.
Mehrtens	"Vorlesungen über Ingenieurwissenschaften" ; Second Edition. Leipzig, 1909 and later.
Merckel	C. Merckel, "Die Ingenieurtechnik im Altertum". Berlin, 1899.
Meyer	A. G. Meyer, "Eisenbauten, Ihre Geschichte und Aesthetik". Esslingen, 1907.
Navier	L. M. H. Navier, "Résumé des Leçons données à l'Ecole des Ponts et Chaussées sur l'Application de la Mécanique à l'Etablissement des Constructions et des Machines" ; New Edition. Brussels, 1839.
Olschki	L. Olschki, "Geschichte der neusprachlichen wissenschaftlichen Literatur" :
	Vol. I. "Die Literatur der Technik und der angewandten Wissenschaften vom Mittelalter bis zur Renaissance" ; Heidelberg, 1919 ;
	Vol. II. "Bildung und Wissenschaft im Zeitalter der Renaissance in Italien" ; published by S. Olschki, 1922 ;
	Vol. III. "Galilei und seine Zeit". Halle, 1927.

Perronet	J. R. Perronet, "Déscription des projets et de la construction des ponts de Neuilli, de Mantes, d'Orléans, etc." Paris, 1788.
Poleni	G. Poleni, "Memorie istoriche della Gran Cupola del Tempio Vaticano". Padua, 1748.
Ritter	Professor Dr. W. Ritter, "Anwendungen der graphischen Statik", four volumes. Raustein, Zürich, 1888–1906.
Rondelet	Rondelet, "L'art de bâtir" ; Italian Edition : "Trattato dell'arte di edificare". Mantua, 1832.
Rühlmann	M. Rühlmann, "Geschichte der technischen Mechanik". Leipzig, 1885.
Saint-Venant	Historical introduction to the Third Edition, edited by Saint-Venant, of Navier's "Résumé des Leçons . . ." Paris, 1864.
Schnabel	F. Schnabel, "Deutsche Geschichte im neunzehnten Jahrhundert" ; Vol. III., "Erfahrungswissenschaften und Technik". Freiburg im Breisgau, 1934.

Steinman and Watson, "Bridges and their Builders". New York, 1941.

Todhunter and Pearson, "A history of the Theory of Elasticity and of the Strength of Materials". Cambridge, 1886–1893.

"Tre mattematici",	Le Seur, Jacquier and Boscowich, "Parere di tre mattematici sopra i danni che si sono trovati nella Cupola di S. Pietro sul fine dell 'Anno 1742".
Vasari	Giorgio Vasari, "Le vite de' più eccellenti pittori, scultori ed architetti", 1550. Salani's Edition. Florence, 1913.
Vitruvius	"De architectura libri decem". Most quotations are taken from the English translation by Joseph Gwilt.

The following periodicals are repeatedly quoted in the text :

Annali dei Lavori Pubblici, Rome.
Annales des ponts et chaussées, Paris (from 1831).
Der Bauingenieur, Berlin (from 1920).
Die Bautechnik, Berlin (from 1923).
Engineering News-Record, New York.
Le Génie Civil, Paris.
L'ingegnere, Milan and Rome.
Schweizerische Bauzeitung, Zürich.
Zeitschrift des Vereins Deutscher Ingenieure.

GENERAL INDEX

251

INDEX OF PERSONAL NAMES[1]

---

[1] The names of certain authors mentioned in footnotes have here been omitted as they are not otherwise concerned with the subject matter of this book.

253

INDEX OF PERSONAL NAMES

INDEX OF PLACE NAMES

INDEX OF PLACE NAMES

257

INDEX OF PLACE NAMES